BASIC CONCEPTS IN CHEMISTRY

Peptides and Proteins

SHAWN DOONAN

University of East London

**WILEY-
INTERSCIENCE**

ROYAL SOCIETY OF CHEMISTRY

For ordering and customer service, call 1-800-CALL-WILEY.

Library of Congress Cataloging-in-Publication Data:
Library of Congress Cataloging-in-Publication Data is available.
ISBN: 0-471-55285-2

Typeset in Great Britain by Wyvern 21, Bristol
Printed and bound by Polestar Wheatons Ltd, Exeter

10 9 8 7 6 5 4 3 2 1

Preface

Life on Earth depends on the chemistry of two classes of macromolecule: nucleic acids and proteins. Nucleic acids carry the genetic information which is passed from parents to progeny, and which provides the blueprint for a particular living organism. That blueprint is essentially a set of instructions for making proteins, and it is on the activities of those proteins that life processes depend. It is, then, no surprise that proteins have attracted enormous scientific interest over the last century or so.

Proteins are constructed on a simple pattern, but that pattern allows for an almost endless diversity of structure and function. They are also very large molecules. For these reasons, the study of proteins has provided a very considerable challenge for the chemist who wishes to determine their structures, and to find out how they work. The importance and magnitude of the task is, perhaps, reflected in the large number of Nobel Prizes for Chemistry that have been awarded for discoveries in protein chemistry; the contributions made by the winners of those prizes are given due prominence here.

This book has been written primarily for students of chemistry, although it is hoped that biological scientists who want to understand something of the chemistry of their subject will also find it useful. For the benefit of those chemistry students who have little background knowledge of biology, I have included brief descriptions of the biological significance of some of the proteins dealt with. These digressions should not be considered as essential reading, but I am firmly of the view that understanding something of the biology adds enormously to the interest of protein chemistry. It might even encourage some readers to become biochemists or molecular biologists!

One word of warning is required. The other books in this series are written using mainly IUPAC names for chemical compounds. Protein chemists do not use that system in the books they write, or in the papers they publish. This might cause confusion for students who choose to read some of the papers and books referred to here. To avoid this difficulty, I have used the nomenclature conventionally adopted in the field, but on first mention of a compound the IUPAC name is also given in parentheses.

This book is dedicated to the memory of Professor Charles Vernon, to whom I owe, besides much else, my interest in protein chemistry.

Shawn Doonan
London

BASIC CONCEPTS IN CHEMISTRY

EDITOR-IN-CHIEF

Professor E W Abel

EXECUTIVE EDITORS

Professor A G Davies
Professor D Phillips
Professor J D Woollins

EDUCATIONAL CONSULTANT

Mr M Berry

This series of books consists of short, single-topic or modular texts, concentrating on the funda-mental areas of chemistry taught in undergraduate science courses. Each book provides a concise account of the basic principles underlying a given subject, embodying an independent-learning philosophy and including worked examples. The one topic, one book approach ensures that the series is adaptable to chemistry courses across a variety of institutions.

TITLES IN THE SERIES

Stereochemistry *D G Morris*
Reactions and Characterization of Solids
 S E Dann
Main Group Chemistry *W Henderson*
d- and f-Block Chemistry *C J Jones*
Structure and Bonding *J Barrett*
Functional Group Chemistry *J R Hanson*
Organotransition Metal Chemistry *A F Hill*
Heterocyclic Chemistry *M Sainsbury*
Atomic Structure and Periodicity *J Barrett*
Thermodynamics and Statistical Mechanics
 J M Seddon & J D Gale
Basic Atomic and Molecular Spectroscopy
 J M Hollas
Organic Synthetic Methods *J R Hanson*
Aromatic Chemistry *J D Hepworth,*
 D R Waring and M J Waring
Quantum Mechanics for Chemists
 D O Hayward
Peptides and Proteins *S Doonan*

Further information about this series is available at www.wiley.com/go/wiley-rsc

Contents

1

The Covalent Structures of Peptides and Proteins

Aims

By the end of this chapter you should understand:

- The importance of proteins in life processes
- The structures of the amino acids and how they are joined together to make proteins
- That some amino acid side chains are hydrophilic and some hydrophobic
- How protein structures are written in shorthand
- The difference between amino acid composition and amino acid sequence of a protein
- Why proteins in solution are electrically charged
- That an extremely large number of different proteins can be constructed from a given number of amino acids

1.1   Introduction: Why Proteins are Interesting

Our genes carry the instructions for making a human being, just as those of the disease-causing bacterium *Helicobacter pylori* (one of the causative agents of duodenal ulcers) carry the instructions for making that, clearly very different, organism. Indeed, one of the great scientific advances of the last few years of the 20th century and the beginning of this was the determination of the complete chemical structures of the genetic material (or genomes) of these and other organisms. So what sort of instructions are they? Genomes are largely a set of instructions for making proteins, and one living organism differs from another in the proteins that it makes.

This obviously means that proteins are the key molecules in the processes of life, and it is now known that virtually all the activities which sustain living organisms are carried out by proteins. The following are a few examples of proteins and what they do:

- **Enzymes** catalyse the reactions of life processes, often increasing their rates by many orders of magnitude
- **Regulatory proteins** control functions such as the expression of genetic information and the balance of the chemical reactions that are going on in a cell at any time
- **Transport proteins** carry other molecules from place to place in the organism. For example, the protein haemoglobin transports oxygen around the blood stream
- **Immunoglobulins** (also known as antibodies) provide a first line of defence against foreign proteins or invading pathogens
- **Muscle proteins** carry out the work of muscular contraction
- **Collagen**, one of the most abundant proteins in the body, provides an essential structural component of skin, bone and tendons
- **Toxins** produced by micro-organisms are often proteins. For example, the protein toxin produced by the food-spoilage organism *Clostridium botulinum* is one of the most toxic molecules known

This list of functions obviously provides a good reason why biologists are interested in proteins, but why should chemists share this interest? The reason is that proteins are extremely complex chemical entities, and analysis of their structures and they ways in which they carry out their particular functions are essentially problems in chemistry. It is no surprise then that, for more than 100 years, chemists have been involved in purifying proteins, analysing their structures, and investigating their modes of action. This book provides an introduction to these areas.

1.2   Amino Acids as the Building Bricks of Proteins

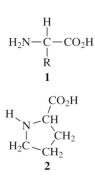

Proteins are made from a set of 19 α-amino acids and the imino acid proline. The general structure of an α-amino acid is shown in **1** and the structure of proline in **2**. Note that in the rest of the book, as in other articles that you might read, the term "amino acid" is often used as shorthand for α-amino acid and is frequently taken to include proline. For example, it is often stated that proteins are made up of 20 different amino acids; this is not strictly correct but is convenient shorthand.

What differentiates one amino acid from another is the nature of the side chain R. Table 1.1 lists the 19 α-amino acids found in proteins according to their chemical types, and gives the structures of the side chains. You need not try to memorize these structures at this stage; they will become familiar as we proceed, and you can look them up in Table 1.1 when necessary.

Table 1.1 The 19 α-amino acids occurring in proteins[a]

Type of side chain	Name	Structure of side chain	Abbreviated name	One-letter abbreviation	Relative molecular mass
Aliphatic	Glycine	—H	Gly	G	75.06
	Alanine	—Me	Ala	A	89.10
	Valine	—CH(Me)Me	Val	V	117.15
	Leucine	—CH$_2$CH(Me)Me	Leu	L	131.17
	Isoleucine	—CH(Me)Et	Ile	I	131.17
Aromatic	Phenylalanine	—CH$_2$—C$_6$H$_5$	Phe	F	165.19
	Tyrosine	—CH$_2$—C$_6$H$_4$—OH	Tyr	Y	181.19
	Tryptophan	—H$_2$C— (indole ring)	Trp	W	204.22
Alcohols	Serine	—CH$_2$OH	Ser	S	105.09
	Threonine	—CH(OH)Me	Thr	T	119.12
Thiol	Cysteine	—CH$_2$SH	Cys	C	121.16
Sulfide	Methionine	—CH$_2$CH$_2$S—Me	Met	M	149.20
Acids	Aspartic acid	—CH$_2$CO$_2$H	Asp	D	133.10
	Glutamic acid	—CH$_2$CH$_2$CO$_2$H	Glu	E	147.13
Amides	Asparagine	—CH$_2$CONH$_2$	Asn	N	132.12
	Glutamine	—CH$_2$CH$_2$CONH$_2$	Gln	Q	146.15
Bases	Lysine	—CH$_2$CH$_2$CH$_2$CH$_2$NH$_2$	Lys	K	146.18
	Arginine	—CH$_2$CH$_2$CH$_2$NH—C(=NH)—NH$_2$	Arg	R	174.20
	Histidine	—H$_2$C— (imidazole ring)	His	H	155.15

[a] The imino acid proline (structure **2**) is also a constituent of proteins. Its abbreviated name and one-letter abbreviation are Pro and P, respectively. Its relative molecular mass is 115.13

The names of the amino acids in Table 1.1 are not systematic. They are, however, the names by which they are universally known. In addition to the name, each amino acid is given a three-letter abbreviation, most of which are the first three letters of the name of the molecule. Exceptions are Asn (asparagine) and Gln (glutamine) to distinguish them from their parent carboxylic acids, Ile (isoleucine) to avoid use of the common term Iso, and Trp (tryptophan) to avoid use of the common word Try. Also listed is a set of one-letter abbreviations. In many cases these are the first letter of the name of the amino acid, but this is not always possible since there are frequently two or more amino acids with the same initial letter. In such cases the most commonly occurring amino acid is given the initial letter (*e.g.* A for alanine), and other letters are used for the rest of the group with the same initial letter (D for aspartic acid, N for asparagine and R for arginine). D and R have been assigned on the somewhat fanciful grounds that aspartic sounds a bit like aspar<u>D</u>ic and arginine could be written as <u>R</u>ginine. There are also fairly obvious reasons for <u>F</u>enylalanine and glu<u>E</u>tamic acid. Use of Y, W, N, Q and K for their respective amino acids is not so obvious. It is, however, important to get used to these symbols for reasons that will become apparent later, when we start to deal with the structures of actual proteins.

1.3   Properties of the Amino Acids

We will not deal here with the chemical properties of the amino acids. The α-amino and α-carboxylic acid groups are mainly of interest in the context of the chemical synthesis of peptides. This is the subject of Chapter 2, and so discussion will be deferred until then. Similarly, the side chains show the usual chemical reactivities in the free amino acids, and these can be found in any standard text on organic chemistry. What is more interesting is the ways in which those reactivities can become modified in proteins, and that will be dealt with in Chapter 5.

1.3.1 Stereochemistry

One obvious feature of the amino acids, with the exception of glycine, is that they all contain at least one stereogenic carbon atom; that is, the molecules are chiral. This topic has been dealt with in detail by Morris[1] and will be considered here only briefly. Structures **3** and **4** show **Fischer projections** of the two forms of an amino acid.

Recall that the conventions are: (a) to draw the carbon chain vertically; (b) to place the carbon of highest oxidation state at the top; (c) the vertical bonds go back into the plane of the page and the horizontal bonds come forward out of the plane. Recall also that if a pair of

$$CO_2H \qquad CO_2H$$

$$H_2N\!\!-\!\!\!\!|\!\!-\!\!H \qquad H\!\!-\!\!\!\!|\!\!-\!\!NH_2$$

$$R \qquad\qquad R$$

3   **4**

Chiral means **handed** and relates to the fact that the mirror images of a chiral molecule cannot be superimposed on one another, just as a right hand and a left hand, which are mirror images, cannot be superimposed. All compounds containing a carbon atom with four different groups attached to it have this property. The α-carbon of amino acids, except glycine, is such an atom. The carbon atom which gives rise to chirality of the molecule is referred to as the **stereogenic** atom or centre. The mirror image forms of the molecule are called **enantiomers**.

groups is switched in a Fischer projection, the configuration is inverted; if two pairs are switched, then the configuration is retained. The corresponding stereo diagrams are shown in **5** (equivalent to **3**) and **6** (equivalent to **4**). The Fischer projection on the left represents an amino acid with the L absolute configuration, and has the amino group on the left, whereas the structure on the right is D. These absolute configurations were assigned by chemical conversions that related them to the configurations of the standard molecule glyceraldehyde (2,3-dihydroxypropanal).

Two of the amino acids, threonine and isoleucine, have two stereogenic centres, and so the possibility of **diastereomeric** forms exists. Fischer projections of the L-forms of threonine and isoleucine are shown in **7** and **8**, respectively. The diastereomers of these molecules do not occur in proteins.

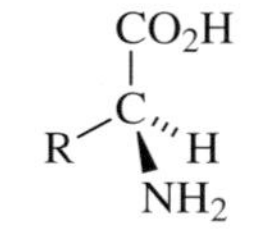

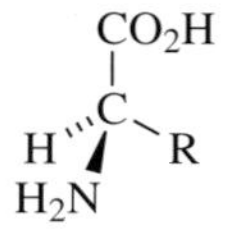

Diastereomers are molecules that differ in arrangement of groups at two or more stereogenic centres but are not enantiomers. L-Threonine (**7**) has a mirror image form (D-threonine) in which the configuration at both stereocentres is reversed. There is also a molecule (**9**) in which the arrangement of groups at the α-carbon is the same as in L-threonine but the configuration at the β-carbon is reversed. This is a different substance with different chemical and physical properties. It is still an L-compound but is called allothreonine. L-Threonine and L-allothreonine are diastereomers.

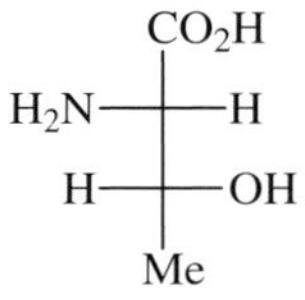

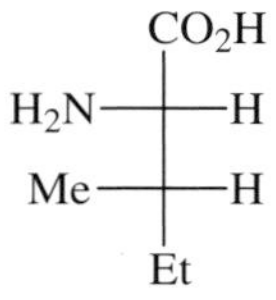

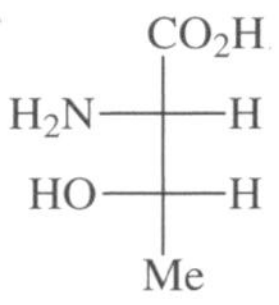

Box 1.1 The Cahn–Ingold–Prelog Convention

Modern practice is to describe the configurations of chiral molecules using the Cahn–Ingold–Prelog *R/S* convention.[2] Under this convention the groups attached to a stereogenic centre are assigned a priority from 1 to 4 in order of decreasing atomic number. The molecule is then viewed with the group of priority 4 behind the stereogenic carbon. If the order of the remaining groups is 1 → 2 → 3 clockwise, then the configuration is *R*. Otherwise it is *S*. When, as is frequently the case, two or more of the atoms directly bonded to the stereogenic centre are carbons, the priority depends on the atomic numbers of the groups next along. Detailed rules are given by Morris.[1] For example, consider L-alanine (structure **5** with R = Me). The priority of the groups attached to the central carbon is NH_2 → CO_2H → Me → H. CO_2H is of greater priority than Me because of the oxygen atoms. Viewing the molecule with the H at the back, the decreasing order of priority of the others is anticlockwise. Hence the configuration is *S*.

In spite of acceptance of the *R/S* convention in nearly all areas of chemistry, the D/L convention is still almost universally used to describe amino acids and carbohydrates. This is unfortunate since it leads to ambiguities. For example, by the convention applied to carbohydrates, L-threonine (**7**) would be a D-compound.

Worked Problem 1.1

Q Structure **10** is the Fischer projection of one enantiomer of cysteine. What is its absolute configuration in both the D/L and *R/S* nomenclatures?

$$CO_2H$$

H——CH_2SH

$$NH_2$$

10

$$CO_2H$$

$$C$$

HSCH_2 H

$$NH_2$$

11

A One way to approach this question is to make a model of **10** and compare it with **5** and **6**. It should be immediately obvious that it has the L configuration. Alternatively, see how many switches of pairs of groups are required to convert **3** into **10**. Switching first the NH_2 and the H, and then the NH_2 and the CH_2SH, yields **10**. An even number of switches retains the conformation. To assign the configuration in the *R/S* notation, first note that the priority order is $NH_2 \rightarrow CH_2SH \rightarrow CO_2H \rightarrow H$ (usually the priority of the side chain is lower than that of the carboxylic acid group, but in this case the sulfur atom changes the order). Draw a stereo diagram of the molecule with the H behind as in **11**. The remaining groups decrease in priority in clockwise order, and thus the molecule is *R*.

One of the fascinating things about the chemistry of living systems is that it seems to be inherently asymmetric. The amino acids found in proteins always have the L configuration. How this came about is a mystery. Most people believe that living organisms arose by chance association of simple molecules present in the oceans of pre-biotic Earth. There is, however, no easy way to explain how it happened that the amino acids selected were entirely of one configuration. The same holds true of virtually all biomolecules that are chiral: one enantiomer is biologically active, the other is not. In the case of the amino acids, small amounts of the D enantiomers do occur in Nature, for example in the cell walls of certain micro-organisms. If these are ingested by animals, enzymes called D-amino acid oxidases convert them into the corresponding aldehydes and thence to carboxylic acids. They are never incorporated into our proteins.

1.3.2 Acid/Base Properties

All amino acids have an α-amino group with a pK_a of about 9–10 and α-carboxylic acid group with a pK_a of about 2–2.5. Hence in neutral solution an amino acid will exist predominantly in the zwitterionic form shown in **12**. These pK_a values are lower than those of monofunctional carboxylic acids and amines because the positive charge on the amino group of the amino acid at low pH stabilizes the negative charge on the carboxylate and *vice versa*. The properties of the amino acids will therefore be pH dependent. For example, the amino group of an amino acid will only be a good nucleophile at pH values greater than about 10.

$$^+NH_3CHCO_2^-$$
$$|$$
$$R$$

12

A **zwitterion** is a molecule that contains both a full positive and a full negative charge. Zwitterions of amino acids are internal salts; that is, the acidic proton of the carboxylic acid function is transferred to the basic amino group.

Box 1.2 Acids, Bases and Acidity

The pK_a is a measure of the strength of an acid. Consider the dissociation of a weak acid HA according to equation (1.1). The equilibrium constant K for the reaction can be written as in equation (1.2):

$$HA + H_2O \rightarrow A^- + H_3O^+ \qquad (1.1)$$

$$K = \frac{[A^-][H_3O^+]}{[HA][H_2O]} \qquad (1.2)$$

However, in dilute solution, $[H_2O]$ is effectively constant and can be incorporated into K to give K_a. Using H^+ as a shorthand for H_3O^+, equation (1.2) becomes equation (1.3):

$$K_a = \frac{[A^-][H^+]}{[HA]} \qquad (1.3)$$

From this it can be seen that the larger is K_a the larger is $[H^+]$, *i.e.* the stronger the acid. It is usual to describe acid strength in terms of the pK_a, which is defined as $-\log K_a$. Because of the minus sign this means that the larger the pK_a the weaker the acid, and as the pK_a value changes by one the acid strength (because of the log relationship) changes by a factor of 10.

Again, the acidity of a solution is usually described by its pH, which is defined as $-\log[H^+]$. Hence, the lower is the pH, the greater the acidity. To a first approximation, a neutral solution has a pH of 7.0. Taking logs of both sides of equation (1.3) and rearranging gives equation (1.4):

$$pH = pK_a + \log \frac{[A^-]}{[HA]} \qquad (1.4)$$

From this it follows that the pK_a is numerically equal to the pH at which the acid is half ionized, *i.e.* where $[A^-]$ is equal to $[HA]$.

Turning to bases, one way to describe their strengths would be in terms of the process described in equation (1.5), and hence of the corresponding K_b (equation 1.6) and pK_b values:

$$B + H_2O \rightarrow BH^+ + OH^- \qquad (1.5)$$

$$K_b = \frac{[BH^+][OH^-]}{[B]} \qquad (1.6)$$

It is, however, more conventional to quote the pK_a of the conjugate acid; that is, to describe the strength of the base in terms of the acid strength of BH^+. These are related by $pK_a + pK_b = 14$.

As we shall see in Section 1.4, the α-amino and α-carboxyl groups of most of the amino acids in a peptide or protein are involved in linking the units together, and so their ionization characteristics are not particularly important in determining the properties of proteins. What is important is that some of the amino acids have ionizable side chains, as can be seen from Table 1.1. Both aspartic acid and glutamic acid have a carboxylic acid group in the side chain. When these amino acids are incorporated into a protein, the pK_a value of the carboxylic acid group is in the range 3.5–4.5. The precise value will depend on the position of the amino acid in the protein, for reasons that will be dealt with in Chapter 5; the important point is that at neutral pH these residues will carry a negative charge. The thiol side chain of the amino acid cysteine is also weakly acidic, with a pK_a value in the region of 9 (more properly, one should say that the thiolate anion is weakly basic). The thiols in proteins are generally uncharged, but in some enzymes, for example, a particular thiol might be in an environment that favours its ionization, thus making it a good nucleophile.

The basic amino acid side chains are those of lysine, arginine and histidine. In proteins the pK_a values of these side chains are in the ranges 9.5–10.5, 12–13 and 6–7, respectively. Hence, in neutral solution, lysine and arginine residues in proteins will be charged. Arginine is a relatively strong base because of resonance stabilization of the protonated guanidinium side chain, and indeed the structure of the amino acid is usually shown in its charged form.

Histidine is interesting in that its pK_a value is close to the pH in living cells. Hence the charge on a particular histidine in a protein will vary with changes in pH around the physiological value (usually taken to be 7.4).

It is important to note that the side chains of the amino acids asparagine and glutamine are not basic. This is because the unprotonated amide group is resonance stabilized, the stabilization being lost if the group is protonated. Similarly, the side chain of tryptophan contains a resonance-stabilized aromatic system and it is only very weakly basic. Under the conditions of pH in which proteins can be studied, the side chain in tryptophan may be considered as non-basic.

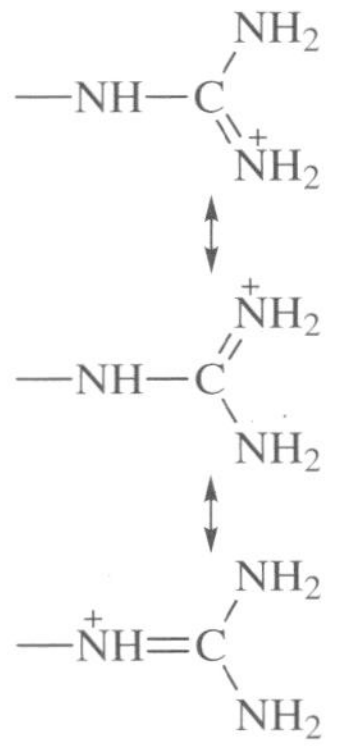

1.3.3 Physical Properties

The amino acids were listed in Table 1.1 according to chemical type. An equally important classification depends on the **hydrophilic/hydrophobic** properties of the side chains; that is, on the affinity that the side chains show for water. This is a property of paramount importance for the way in which proteins fold up into compact three-dimensional structures, and will be returned to in Chapter 5.

| **Box 1.3   Hydrophobicity** |

Hydrophobicity is quite a subtle concept, although its manifestations are often obvious. For example, if a little olive oil, which is a mixture of triglycerides, is poured into water then the oil does not dissolve but rather forms droplets. This is explained in terms of "hydrophobic bonding" between the non-polar triglyceride molecules, and could be thought to show that the triglycerides interact favourably with one another but not with water molecules; that is, the explanation is essentially one concerned with the enthalpies of interaction. In fact this is not correct; the effect is essentially entropic. A triglyceride molecule in water is surrounded by an ordered layer of water molecules. If the triglyceride is removed from water into a lipid droplet, then the constraints on the water molecules will be removed, resulting in an increase in the entropy of the system. It is essentially this increase in entropy that drives the formation of so-called "hydrophobic bonds". A detailed analysis of the hydrophobic effect has been given by Tanford.[3]

It is not easy to quantify the hydrophilic/hydrophobic properties of the amino acids. A variety of quantitative measurements have been made of the partition of derivatives of the side chains or of the amino acids between water and organic solvents to obtain scales of hydrophobicity,

The three **resonance** forms of the guanidinium part of the side chain of arginine are shown below. The actual state of the group is a **hybrid** of these, with the proton shared between the three nitrogen atoms. The hybrid is more stable than any one of the three individual forms. Loss of a proton would result in the loss of resonance stabilization and hence guanidine is quite a strong base and the guanidinium ion is a very weak acid.

but the values obtained vary somewhat depending on the precise system used. The results of one such set of experiments are given in Table 1.2. Here the partition of *N*-acetyl (*N*-ethanoyl) amino acid amides ($CH_3CONHCHRCONH_2$) between water and octanol was determined;[4] the results are presented as free energies of transfer of the compound from octanol to water and normalized to zero for glycine. The larger the positive value, the greater the hydrophobicity; the larger the negative number, the greater the hydrophilicity. On this basis, the amino acids fall into three groups that merge into one another. The acidic and basic amino acids (except histidine) and the amides are hydrophilic. The hydroxy-amino acids, glycine, histidine and alanine, are intermediate in properties. The aliphatic and aromatic amino acids, proline and methionine, are hydrophobic.

The values in the final column of Table 1.2 give the so-called hydropathy index of the amino acids.[5] This is a different type of scale in which

Table 1.2 The hydrophilic/hydrophobic nature of the amino acids

Amino acid	Free energy of transfer of N-acetyl amino acid amides from octanol to water (kJ mol^{-1})	Hydropathy index[a]
Tryptophan	9.41	−0.9
Isoleucine	7.52	4.5
Phenylalanine	7.48	2.8
Leucine	7.11	3.8
Cysteine	6.43	2.5
Methionine	5.14	1.9
Valine	5.10	4.2
Tyrosine	4.01	−1.3
Proline	3.00	−1.6
Alanine	1.29	1.8
Threonine	1.09	−0.7
Histidine	0.54	−3.2
Glycine	0	−0.4
Serine	−0.17	−0.8
Glutamine	−0.92	−3.5
Asparagine	−2.54	−3.5
Glutamic acid	−2.67	−3.5
Aspartic acid	−3.22	−3.5
Lysine	−4.14	−3.9
Arginine	−4.22	−4.5

[a] A scale in which experimental values of free energies of transfer are combined with information obtained from the locations of amino acid residues in proteins to give a consensus scale of relative hydrophilicity/hydrophobicity

experimental values of free energies of transfer are combined with information obtained from the locations of amino acid residues in three-dimensional structures of proteins (see Chapter 5) to give a consensus scale of relative hydrophilicity/hydrophobicity. It can be seen that the relative positions of the amino acids are now rather different when compared with the values for free energy of transfer. In particular, tryptophan and tyrosine are moderately hydrophilic on this scale, and histidine quite strongly so. This reflects the common occurrence of these amino acids on the surfaces of proteins, where they play special roles. Nevertheless, the broad outlines of the two classifications are similar and help us to understand, in most cases, the locations in which we find the various amino acids in the three-dimensional structures of proteins.

1.4   The Peptide Bond

Amino acids are the structural units of peptides and proteins, but how are these units – or **residues** as they are termed – linked together? As long ago as 1902, Emil Fischer noted that proteins contain relatively few free amino and carboxyl groups. On this basis he suggested that the linkage between them involves condensation of the carboxylic acid group of one amino acid with the amino group of the next to form an amide linkage, as shown in Scheme 1.1. The linkage produced is referred to as the **peptide bond**. The product of Scheme 1.1 is known as a **dipeptide**. Simply repeating this process leads to the generalized structure of a **protein** as shown in **13**, where the R groups represent the side chains of any of the amino acids in Table 1.1.

Fischer also made important contributions to ideas about how enzymes catalyse reactions. He was awarded the Nobel Prize for Chemistry in 1902, but this was for earlier work that he had done on the biosynthesis of sugars and purines. He was obviously a chemist with wide interests!

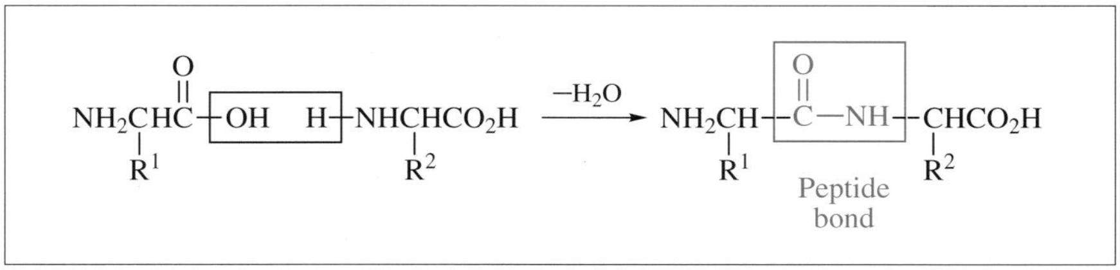

Scheme 1.1

$$NH_2CHCO-[NHCHCO]_n-NHCHCO_2H$$

13

Note that peptide bond synthesis in Nature does not occur like this. The interested reader can consult any standard textbook of biochemistry to find out how Nature does it.

The two ends of a protein are different. That on the left of **13** has a free α-amino group and is termed the **N-terminus**, whereas that on the right has a free carboxylic acid group and is termed the **C-terminus**. *By* *convention, protein structures are always written this way around.* This becomes important when we use shorthand representations of protein structures as described below.

Recall that two of the amino acids contain carboxylic acid groups in their side chains (aspartic and glutamic acids) whereas one, lysine, contains an amino group. The possibility therefore arises that proteins could contain branches, with chains growing from the side chain carboxylic acid or amino groups. They never do: the way in which proteins are biosynthesized makes it impossible. This is not to say that branched peptides never occur. The biologically active peptide glutathione, or γ-glutamylcysteinylglycine (**14**), has a peptide bond to the side chain of glutamic acid, but this peptide is made by a route different from that used to synthesize proteins.

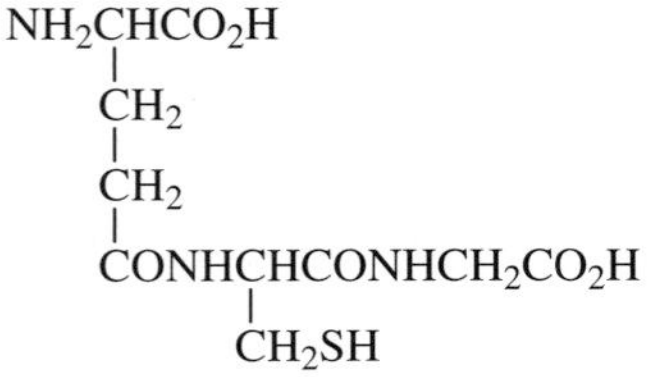
NH2CHCO2H
CH2
CH2
CONHCHCONHCH2CO2H
CH2SH

14

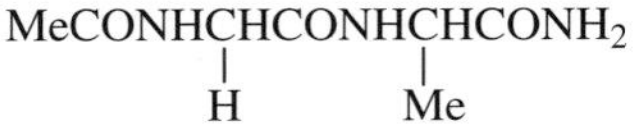
MeCONHCHCONHCHCONH2
H
Me

15

Just as lactones are cyclic esters, **lactams** are cyclic amides; that is, they are formed by internal elimination of water from molecules containing both a carboxyl group and an amino group. For example, **16** is formed by elimination of water from $NH_2CH_2CH_2CH_2CH_2CO_2H$. It is called δ-valerolactam or 2-azacyclohexanone.

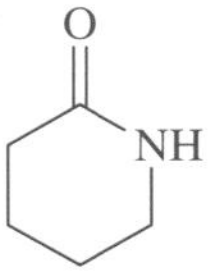
O
NH

16

Other variations on the usual theme also occur, in particular modification of either the N-terminal amino group or C-terminal carboxyl group, or both. Frequently found modifications are acetylation (ethanoylation) of the N-terminus and amidation of the C-terminus. Structure **15** shows a peptide with both of these modifications. Another possibility for blocking the N-terminus when the terminal residue is glutamine involves a condensation reaction between the amino group and the side chain amide to form a lactam (Scheme 1.2). The resulting residue is known as pyroglutamic acid. Modifications at the termini are probably introduced to prolong the activity of peptides in the body by preventing the action of enzymes that catalyse the hydrolysis of peptides at the terminal residues.

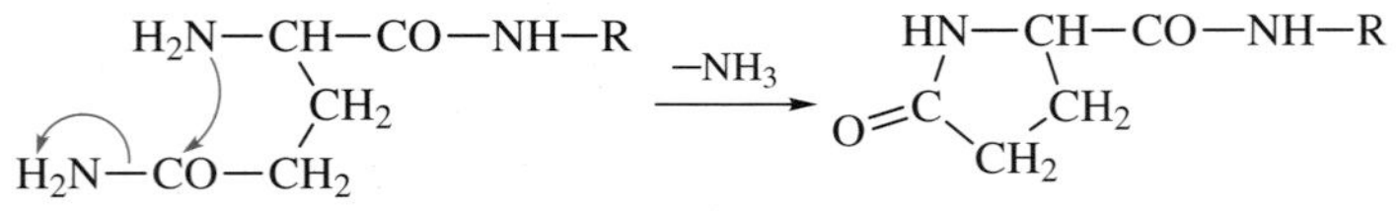
H2N—CH—CO—NH—R
CH2
H2N—CO—CH2
−NH3
HN—CH—CO—NH—R
O=C
CH2
CH2

Scheme 1.2

Worked Problem 1.2

Q The peptide thyrotropin releasing factor (TRF), which stimulates the release of a peptide hormone called thyrotropin from the pituitary gland, has the structure pyroglutamylhistidinylproline amide. Give its covalent structure (it is unnecessary to specify states of ionization of the residues, nor their stereochemistry).

A This problem is complicated because both the N- and the C-termini of the peptide are blocked. The N-terminus is pyroglutamate (see Scheme 1.2) and the C-terminus is amidated (see **15**). The structure of the peptide is shown in **17**.

17

One other important modification which occurs in some peptides and proteins involves the amino acid cysteine. In solution, under mildly oxidizing conditions, two molecules of cysteine react to give a **disulfide bridge** and produce the amino acid cystine (**18**). The same thing can happen when a peptide or protein contains two residues of cysteine. The disulfide bridge in this case links together two, possibly remote, regions of the polypeptide chain that are brought together by the way in which the chain folds up in space. Some proteins are known that contain several such disulfide bridges. A few proteins consist of two or more polypeptide chains linked together by disulfide bridges. One example is insulin, whose structure is described in Section 1.9.

It is important to note that this is not a chance occurrence. Many peptides and proteins contain pairs of cysteine residues that do not produce disulfide bridges. Where two or more pairs of cysteines in a particular protein molecule form bridges, they always link in precisely the same way; it is part of the architecture of the protein, and the links are introduced after the protein has been synthesized.

Cysteine and cystine are written very similarly and are easy to confuse. The names are, however, pronounced quite differently. Cysteine is pronounced with three syllables as "sis-tay-een" whereas cystine is pronounced as "sys-teen".

18

1.5   Terminology and Shorthands

The product of Scheme 1.1 was termed a dipeptide. The structure in **13** is a protein. The difference is only one of size and the point at which the terminology changes is not clear-cut. Molecules containing up to 20 residues would certainly be referred to as peptides (or sometimes as polypeptides or oligopeptides). Molecules containing 50 residues would usually be considered as proteins. Where the switch occurs is not particularly important. What is important is that chemically there is no distinction between the two: peptides are small proteins.

Structure **19** represents a pentapeptide. It can be named as glycyllysylserylaspartylalanine. It is clearly laborious to draw structures such as **19** and becomes rapidly more so, or indeed impossible, as the chain length increases. Hence, for many purposes, shorthand is used where the structure is represented by a string of the three-letter abbreviations for the amino acids. To a protein chemist the string of abbreviations Gly-Lys-Ser-Asp-Ala means just the same as structure **19**. It is, of course, essential to recall the convention that the amino acid on the left is the N-terminus and that on the right is the C-terminus. The pentapeptide Ala-Asp-Ser-Lys-Gly is a different molecule.

$$NH_2CH_2CONHCHCONHCHCONHCHCONHCHCO_2H$$

$$
\begin{array}{cccc}
| & | & | & | \\
CH_2 & CH_2OH & CH_2 & Me \\
| & & | & \\
CH_2 & & CO_2H & \\
| & & & \\
CH_2 & & & \\
| & & & \\
CH_2NH_2 & & &
\end{array}
$$

19

For some purposes, and in particular when it is required to store and analyse large amounts of data about protein structures (see Chapter 6), even this shorthand is not short enough. It was for this reason that the single letter code (Table 1.1) was introduced. In its most abbreviated form, structure **19** may be represented as GKSDA; to a protein chemist this string of five letters has the same information content as structure **19**.

1.6   Composition and Sequence

The peptide in **19** contains one residue each of the amino acids alanine, aspartic acid, glycine, lysine and serine. That is its amino acid composition. There are 120 ways in which one residue of each of these five amino acids can be linked together to form a pentapeptide. Each of these represents a different amino acid sequence (also referred to as primary structure). So the amino acid composition of a peptide or protein is the number of residues of each of the constituent amino acids that the molecule contains. The amino acid sequence (primary structure) describes the order in which these amino acids occur in the polypeptide chain.

Worked Problem 1.3

Q There are two possible dipeptides containing one residue each of aspartic acid and lysine. Draw their structures. Write down their structures in shorthand, using both the three- and the one-letter abbreviations for the residues.

A We are given the amino acid composition of the two peptides. What we want is their amino acid sequences. The structures of the side chains are given in Table 1.1 and the residues are linked by a peptide bond so the structures are as shown in **20** and **21**.

$$
\begin{array}{cc}
\text{NH}_2\text{CHCONHCHCO}_2\text{H} & \text{NH}_2\text{CHCONHCHCO}_2\text{H} \\
\;\;|\qquad\quad| & \;\;|\qquad\quad| \\
\text{CH}_2\quad\;\text{CH}_2 & \text{CH}_2\quad\;\text{CH}_2 \\
\;\;|\qquad\quad| & \;\;|\qquad\quad| \\
\text{CO}_2\text{H}\quad\text{CH}_2 & \text{CH}_2\quad\text{CO}_2\text{H} \\
\qquad\quad| & \;\;| \\
\qquad\;\text{CH}_2 & \text{CH}_2 \\
\qquad\quad| & \;\;| \\
\qquad\text{CH}_2\text{NH}_2 & \text{CH}_2\text{NH}_2 \\
\textbf{20} & \textbf{21}
\end{array}
$$

In three-letter abbreviations the two sequences are Asp-Lys and Lys-Asp, respectively, and in the one-letter abbreviations DK and KD, respectively, for **20** and **21** (remember that the convention is to write the N-terminal residue on the left).

Worked Problem 1.4

Q Would a peptide with the amino acid sequence KRYDEQN be hydrophobic or hydrophilic in character?

A To answer this it will be necessary to look up the hydrophobicity values for the residues in Table 1.2. There the amino acids are listed under their full names, so we need to translate the one-letter abbreviations in order to look them up. They are K = lysine, R = arginine, Y = tyrosine, D = aspartic acid, E = glutamic acid, Q = glutamine and N = asparagine. With the exception of tyrosine, these are all very hydrophilic amino acids. Hence the peptide will have a strongly hydrophilic character.

1.7   Electrical Charges on Peptides and Proteins

Structure **19** has been drawn without specifying the charges on the amino acids; this is usual practice. Peptides and proteins are almost invariably studied in aqueous solution at a particular pH, and sometimes it is necessary to focus on the charge that the molecule will have. From the approximate pK_a values given in Section 1.3.2 it is easy to predict what the approximate charge state of the molecule will be at extremes of pH. At pH values less than 3, both amino groups will be protonated to give $-NH_3^+$, as will both carboxyl groups to give $-CO_2H$; that is, the molecule will carry a charge of 2+. At high pH, say above 10, both amino groups will be uncharged ($-NH_2$) and both carboxylate groups ionized ($-CO_2^-$) and the net charge will be 2–. At intermediate pH values around 7, all four groups will be ionized. This means that the net charge will be zero because the peptide carries an equal number of positive and negative charges. For any peptide or protein there will be some value of the pH at which the net charge is zero (*i.e.* the numbers of positive and negative charges are equal). This is referred to as the isoelectric pH or, more commonly, the isoelectric point. The variation of charge on a protein with pH will be important to us for the discussion of protein purification in Chapter 3.

1.8   An (Almost) Infinite Variety of Structures

As outlined in the Section 1.1, peptides and proteins perform a vast range of biological functions, but we have just seen that their structures are based on a rather simple pattern: amino acids linked together by peptide bonds. There is no contradiction here. Consider a decapeptide, that is, a molecule containing 10 amino acids linked in a chain. Given that there are 20 different amino acids, how many such peptides could in principle exist? The first residue could be any one of 20, as could the second. So there are 20×20 possible dipeptides. Each of these 20^2 dipeptides could be converted to 20^3 tripeptides, and so on. There are hence 20^{10} or 10,240,000,000,000 possible decapeptides that could be made from the 20 amino acids listed in Table 1.1. This is a very large number indeed, and the possibilities increase very rapidly as the length of the chain increases. Given also that the amino acids have a range of different chemical and physical types, it is not surprising that Nature has been able to construct proteins to carry out so many diverse functions.

1.9   An Example of Protein Structure: Insulin

The hormone insulin was mentioned in Section 1.4 as an example of a protein whose structure consists of two polypeptide chains linked by disulfide bridges. The structure of human insulin is shown in **22** using the three-letter code for the amino acids. Insulin is a very small protein

---

Worked Problem 1.5

Q How many different tripepetides could be made with the amino acid composition Gly, Ala, Val? Write their amino acid sequences using the one-letter abbreviations.

A There are three ways of choosing the N-terminal residue. For each of these there is a choice of two amino acids for the second residue. After that there is only one left. So the number of peptides is $3 \times 2 \times 1 = 6$. Their sequences are:

GAV   GVA   AGV   AVG   VAG   VGA

– indeed it is frequently referred to as a peptide hormone – but it should be clear that it is only practical to write down its structure using the shorthand representations for the amino acids. The two chains of insulin are called the A- and B-chains. The A-chain contains 21 amino acid residues and has a glycine at its N-terminus; it also has a disulfide bridge linking cysteines 6 and 11. The B-chain contains 30 amino acids and has phenylalanine as the N-terminal residue. It does not contain any internal disulfide bridges, but is linked to the A-chain by two bridges joining residue B7 to A7 and residue B19 to A20. If you are wondering how the two chains of insulin "find" one another after synthesis in the cell, the answer is that they do not. Insulin is synthesized as an inactive precursor; that is, as a larger, single-chain protein from which an internal section is subsequently removed to produce the two-chain active hormone. The bridges are formed before the internal section is removed.

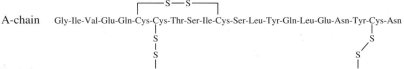

A-chain   Gly-Ile-Val-Glu-Gln-Cys-Cys-Thr-Ser-Ile-Cys-Ser-Leu-Tyr-Gln-Leu-Glu-Asn-Tyr-Cys-Asn

B-chain   Phe-Val-Asn-Gln-His-Leu-Cys-Gly-Ser-His-Leu-Val-Glu-Ala-Leu-Tyr-Leu-Val-Cys-Gly-Glu-Arg-Gly-Phe-Phe-Tyr-Thr-Pro-Lys-Thr

22

Summary of Key Points

- Proteins are the chemical devices that carry out virtually all of the functions of living organisms. Understanding their structures and how they work are essentially chemical problems.
- Proteins are macromolecules made from up to 19 different amino acids and the imino acid proline, linked together by peptide bonds.

- The amino acids, with the exception of glycine, are chiral molecules, but only one enantiomer (the L-form) occurs in proteins.
- Several of the amino acids contain ionizable side chains so that a protein in solution will carry electrical charges. At one particular pH, the isoelectric point, the numbers of positive and negative charges will be equal and the net charge will be zero.
- Some of the amino acid side chains are hydrophobic and some are hydrophilic. This has important consequences for the way in which polypeptide chains fold up in space.
- The difference between peptides and proteins is simply one of size: peptides are small proteins.
- The amino acid composition of a peptide or protein is a list of which amino acid residues the molecule contains and how many copies there is of each. The amino acid sequence is the order in which the amino acid residues occur in the polypeptide chain.
- Peptides and proteins have two unique ends. At one end, termed the N-terminus, there is a free α-amino group. At the other, termed the C-terminus, there is a free α-carboxyl group. Conventionally structures are written with the N-terminus on the left. In some peptides and proteins the terminal groups may be chemically modified.
- Each amino acid can be represented by both a three-letter and a one-letter abbreviation. Peptide and protein structures can be written in shorthand as a string of these abbreviations, but it is essential to retain the convention that the residue on the left is N-terminal.
- Because of the large number of sequences that can be constructed for a given amino acid composition, and because the amino acids show a wide variety of chemical and physical characteristics, Nature has been able to construct proteins to carry out any required biological function.

Problems

1.1. Assign *R/S* configurations to both stereocentres in structure **7** (L-threonine).

1.2. Explain, on the basis of resonance structures, why the side chain of the amino acid glutamine is not basic.

1.3. There are two possible dipeptides containing one residue each of tryptophan and asparagine. Draw their structures. Write down their structures in shorthand using both the three- and the one-letter

1.4. The artificial sweetener aspartame is aspartylphenylalanyl methyl ester. Draw its structure (both amino acids have the L configuration, but it is not necessary to specify this in the structure that you draw).

1.5. What would be the net charge on the peptides in Problem 1.3 at pH 2 and pH 12?

1.6. How many different tetrapepetides could be made with the amino acid composition Asp, Trp, Ile and Phe? Write their amino acid sequences using the one-letter abbreviations.

1.7. What is the amino acid composition of the peptide ADAGFWKFAAGPS? Re-write the sequence using three-letter abbreviations.

1.8. How many possible proteins each containing 50 amino acid residues could be constructed from the 19 amino acids plus proline? Assume that the average relative molecular mass of a residue is 110. What would be the average relative molecular mass of each of these proteins? Assuming that you made one molecule of each of these proteins, what would be the total mass (take an approximate value of 6×10^{23} for the Avogadro constant)?

References

1. D. G. Morris, *Stereochemistry*, The Royal Society of Chemistry, Cambridge, 2001, p. 23.
2. R. S. Cahn, C. K. Ingold and V. Prelog, *Angew. Chem. Int. Ed. Engl.,* 1966, **5**, 385.
3. C. Tanford, *The Hydrophobic Effect: Formation of Micelles and Biological Membranes*, Wiley-Interscience, New York, 1973.
4. J. Fauchère and V. Pliska, *Eur. J. Med. Chem.,* 1983, **18**, 369.
5. J. Kyte and R. F. Doolittle, *J. Mol. Biol.,* 1982, **157**, 105.

Further Reading

A. L. Lehninger, D. L. Nelson and M. M. Cox, *Principles of Biochemistry,* 2nd edn., Worth, New York, 1993, pp. 111, 892.
N. J. Darby and T. E. Creighton, *Protein Structure*, IRL Press, Oxford, 1993, p. 1.
M. Jones, *Organic Chemistry,* Norton, New York, 1997, p.1343.

2
Chemical Synthesis of Peptides

Aims

By the end of this chapter you should understand:

- Some of the reasons for the chemical synthesis of peptides
- The general principles underlying peptide synthesis and the importance of protecting groups
- The benefits and difficulties of solid-phase peptide synthesis
- How peptide bonds are made
- How peptides are linked to the solid support in solid-phase peptide synthesis
- How the nature of the N-terminal protecting group influences the choice of side-chain protecting groups
- How synthetic peptides are purified and characterized
- The conditions under which racemization can occur and how it can be avoided

Hormones are substances produced by specialized structures called **endocrine glands**. Hormones are released into the blood stream and carried to their target organs, where they exert a stimulatory effect on a specific biochemical or physiological process. For example, **growth hormone**, which is made in the **pituitary gland**, stimulates the growth of the long bones. Its absence in the early stages of development leads to dwarfism. Note that not all hormones are polypeptides. The **sex hormones**, for example, are steroids.

2.1 Introduction

Peptides have a wide range of interesting and important biological activities. A few examples may help to give an idea of the range of their functions. The level of glucose in the blood is regulated by the two peptide **hormones** insulin and glucagon; insulin decreases the level of blood glucose by increasing its uptake into the liver and its storage as glycogen, whereas glucagon has the opposite effect of promoting the breakdown of glycogen. The nine-residue peptide oxytocin (**1**) has the effect of stimulating uterine contractions during childbirth. Note that oxytocin has a disulfide bridge linking residues 1 and 5 and is amidated at the C-terminus. Structures **2** and **3** are a pair of peptides called enkephalins that are produced in the central nervous system and have the effect of decreasing the sensation of pain; they are Nature's painkillers and in many ways mimic the action of morphine.

$$\overline{S-S}$$
Cys-Tyr-Ile-Gln-Asn-Cys-Pro-Leu-GlyNH$_2$

1

Tyr-Gly-Gly-Phe-Met   Tyr-Gly-Gly-Phe-Leu

2   **3**

A question that immediately arises is: how do these peptides work? One way of approaching this question is to synthesize variants or **analogues** where one or more of the amino acid residues is replaced by another, perhaps a non-naturally occurring one, and to investigate the effect of the change on biological activity. Such **structure–activity** studies are a fundamental part of physiology and pharmacology.

The second interesting question is: can we use these biologically active peptides, or analogues of them, therapeutically? The answer is often yes. For example, oxytocin is widely used to induce labour in cases of difficult childbirth and, of course, insulin has been used for many years in the treatment of diabetes. Indeed, there are now many peptides that are routinely used in human health care. The vast majority of these are synthetic, and peptide synthesis is a major activity in pharmaceutical companies just as it is in many academic research laboratories.

2.2   Principles of Peptide Synthesis

It is easy to make a peptide bond (that is, an amide) by reaction of a suitable carboxylic acid derivative with an amine. An example is shown in Scheme 2.1, where the derivative is an acyl chloride. However, there is much more to peptide synthesis than that. If we are trying to make a dipeptide, then mixing the acyl chloride of amino acid 1 (A$_1$) with amino acid 2 (A$_2$) will indeed yield some of the desired dipeptide (A$_1$A$_2$), but we will also get the product of A$_1$ reacting with itself (A$_1$A$_1$) and a complex mixture of larger molecules (A$_1$A$_2$A$_1$, *etc.*). In addition, if one of the amino acids has a nucleophilic side chain, then reaction will also occur at the side chain to yield branched products. The result will be a mess.

Fundamental to peptide synthesis is the use of **protecting groups**; that is, all reactive groups not required to participate in the reaction must be chemically modified to prevent their reaction.

$$R-\overset{\overset{\text{O}}{\|}}{C}-Cl + NH_2-R' \xrightarrow{-HCl} R-\overset{\overset{\text{O}}{\|}}{C}-NHR'$$

Scheme 2.1

A generalized scheme for peptide synthesis is shown in Scheme 2.2. The process is started with an amino acid that has the amino group protected with a group P_1 and, if necessary, the side chain protected with a group P_2. In reaction (i) the carboxyl group of this amino acid is converted into a derivative that will be reactive to nucleophilic attack. The second amino acid has a protected carboxylic acid group (P_3) and, again if necessary, a protected side chain (P_4). Reaction between these two [reaction (ii)] will yield the fully protected dipeptide. There are now two possibilities. Protecting group P_1 can be removed [reaction (iii)]. Addition of another amino-protected carboxyl-activated amino acid will then extend the polypeptide chain at the N-terminus [reaction (iv)]. Alternatively, P_3 can be removed [reaction (v)] to yield a free carboxylic acid group that can be activated and reacted with another carboxyl-protected amino acid to extend the polypeptide at the C-terminus [reaction (vi)]. For reasons that will be explained in Section 2.8, elongation at the N-terminus is the favoured route.

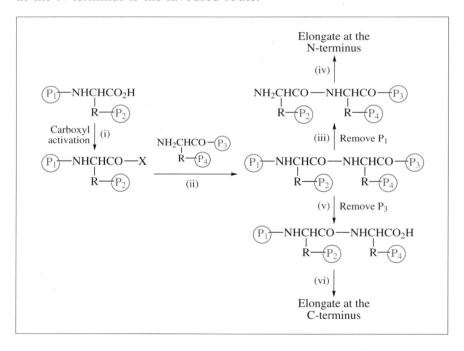

Scheme 2.2

The process in Scheme 2.2 poses some interesting chemical problems. Firstly, there is obviously a need for protecting groups that can be selectively removed during the course of the synthesis. For example, in step (iii) the amino protecting group P_1 must be removed without affecting the other three and without breaking the newly formed peptide bond. In fact, there is a need for two different amino protecting groups if one of the amino acids involved in the synthesis is lysine, because lysine has an amino group in its side chain (see Table 1.1) which will have to remain protected whilst the α-amino group is unprotected. At the end of the synthetic process, all side-chain protecting groups must be removed to yield the final product.

The other major requirement is that the peptide bond formation should be as complete as possible, particularly if it is intended to synthesize a peptide more than a few amino acids long. The reason for this is that the overall yield decreases rapidly with the number of steps in the synthetic process unless the step yield is high. This is illustrated in Figure 2.1, which shows how the overall yield decreases with the number of synthetic cycles for step yields of 90% and 99%. For moderately long peptides (25 residues), a step yield of 90% is clearly unacceptable since it leads to a final yield of less than 10%. Even with a repetitive yield of 99% the overall yield is only about 80%. Obtaining a step yield of 99% obviously imposes great demands on the chemistry of the process, and unrealistic demands on the skill of the chemist carrying it out!

This is a general problem. In any synthetic process with a large number of steps the overall yield decreases rapidly with decreasing step yield.

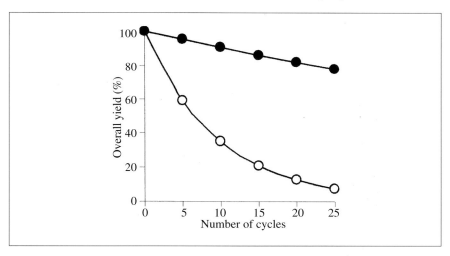

Figure 2.1 Dependence of overall yield on step yield. Open circles: step yield 90%; closed circles: step yield 99%

Worked Problem 2.1

Q What would be the overall yield in the synthesis of a 25-residue peptide if the step yield was 95%?

A Twenty four coupling steps are involved. Hence the fractional yield will be 0.95^{24} and the percentage yield $100 \times (0.95^{24})$. This is 29.2%.

2.3   Solution and Solid-phase Approaches to Synthesis

The original approach to peptide synthesis was to carry out the chemistry in solution. At each stage the product peptide had to be isolated and purified, which required considerable technical skill and resulted in loss of yield even if the coupling step was very efficient. Clearly this approach could only be used to synthesize small peptides in acceptable yield. To extend the method to the practical synthesis of large peptides a process of **block synthesis** followed by **fragment condensation** was developed. In this process the desired peptide is broken down into convenient sized blocks of 5–10 residues. Each of these blocks is synthesized in protected form and then, after appropriate deprotection, the units are sequentially linked to form the final product. The process is shown in outline in Scheme 2.3.

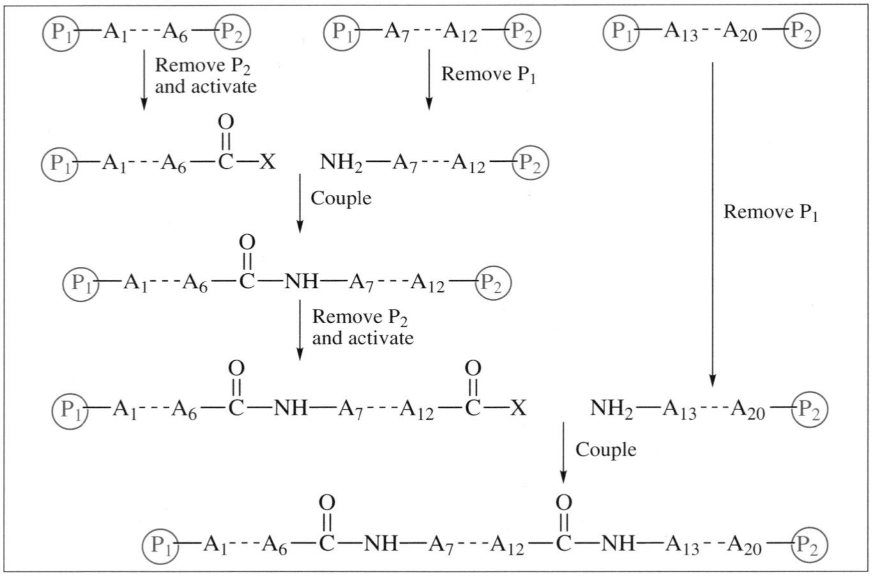

Scheme 2.3

The amino acid residues are shown as A_1, A_2, *etc.*, and, for convenience, condensation of only three blocks of sequence is shown. The final step in the synthesis would be removal of the N- and C-terminal protecting groups and also of those from any side chains that needed protection.

A completely new approach was introduced by Merrifield[1] in the early 1960s. This was **solid-phase peptide synthesis (SPPS)**. In this technique the amino acid that is to be at the C-terminus of the final peptide is N-protected and then attached *via* its carboxyl group to an insoluble polymeric material, usually polystyrene [poly(phenylethene)].

For this contribution, Merrifield was awarded the 1984 Nobel Prize in Chemistry.

Box 2.1   Polystyrene

The most commonly used solid support is a co-polymer of styrene and divinylbenzene. The divinylbenzene provides cross-links between the polymer chains (Scheme 2.4) and the degree of cross-linking can be varied by varying the amount of divinylbenzene in the polymerization reaction. For peptide synthesis a loosely cross-linked polymer is used (1–2%) so that the solvents and reagents used can freely penetrate the polymer beads. The solvent occupies about 90% of the swollen beads and synthesis effectively takes place in free solution even though the peptide chain is linked to the support.

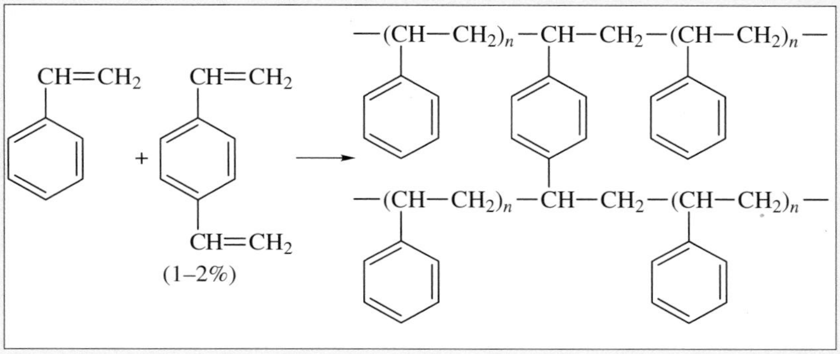

$$-(CH-CH_2)_n-CH-CH_2-(CH-CH_2)_n-$$

$$CH=CH_2 \qquad CH=CH_2$$

$$+$$

$$CH=CH_2$$
$$(1-2\%)$$

$$-(CH-CH_2)_n-CH-CH_2-(CH-CH_2)_n-$$

Scheme 2.4

Linkage to the **solid support** must be such that it is not broken during the course of subsequent peptide synthesis but can be cleaved at the end of the process to liberate the product peptide. The protecting group of the amino acid linked to the solid support is then removed and reacted with the N-protected and carboxyl-activated derivative of the next residue in the chain, and so on. That is, the peptide is built up attached to the solid support from the C-terminus to the N-terminus. The first few steps in the process are shown in outline in Scheme 2.5. The solid support is represented by the brown ball. Protecting groups are also shown in brown with the peptide in black. It is usual practise to draw these schemes from right to left to emphasize that synthesis is carried out from the C-terminus to the N-terminus.

There are some very great advantages in the solid-phase approach to peptide synthesis. The most obvious one is that removal of side products in the reaction is achieved simply by washing the resin on a filter. The growing chain at all times remains attached to the solid support and so does not have to be isolated. More importantly, whereas solubility of protected peptides is often a serious problem in solution phase synthesis, the issue does not arise with SPPS since the peptide is attached to

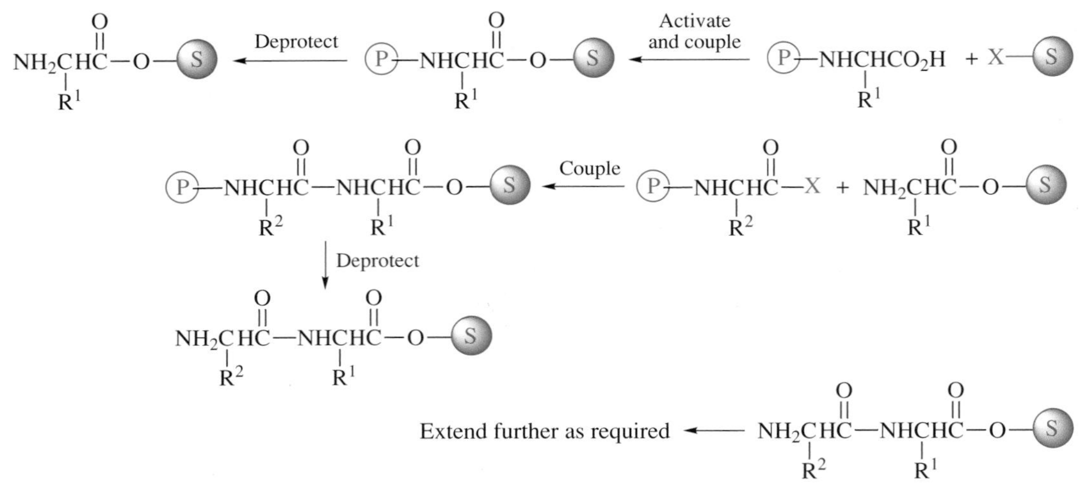

Scheme 2.5

The solid phase method has now been applied in other areas of synthesis. A particularly important application is in the synthesis of **oligonucleotides**, that is, sections of DNA molecules. This technology is central to the methods by which the structures of genomes are determined. Again, oligonucleotide synthesis has been automated.

the support. For these reasons, SPPS is a fairly straightforward process which can be implemented in any research laboratory where there is an interest in peptides. Indeed, the process has been fully automated so that it is now possible to buy a machine which does the work for you. Solution synthesis, on the other hand, is technically difficult and requires considerable experience to be carried out effectively. This is not to say that solution synthesis is no longer practised. It is still widely used in the pharmaceutical industry, particularly for those peptide products that were developed before SPPS became an established method. One of the reasons for this is that if a switch were made to synthesizing licensed products by a different method, the regulatory agencies would require re-validation of the product; this is a long and expensive task. Solution synthesis is also often still used in block synthesis (Scheme 2.3) of large peptides for producing the individual fragments to be linked together. However, because of its more widespread use, SPPS is the main focus of the rest of this chapter.

There is often a down side to technical advances and in the case of SPPS it arises from the very fact that the growing peptide chain remains linked to the solid support. This means that if for some reason the coupling reaction does not go to completion at any step, then the final product will contain a population of molecules with a residue missing. It may be quite difficult to remove this **deletion product**. For this reason it is common practice to check for completion of coupling at each step so that a repeat coupling can be carried out if necessary. This can be done by removing a small sample of the resin and testing for free residual amino groups using the ninhydrin reaction (see Chapter 4). If the repeat coupling still does not proceed to completion, the uncoupled peptide molecules can be permanently blocked to prevent them growing any further (see Section 2.9).

Worked Problem 2.2

Q Suppose that in the synthesis of a pentapeptide (five amino acid residues) by the solid-phase method the second coupling step proceeds only to 80% completion but at all other steps the coupling is complete. What products would be formed and in what yield?

A Suppose that the desired peptide has the sequence A_1–A_2–A_3–A_4–A_5. Synthesis is from the C-terminal end so the second coupling is the reaction which adds A_3 to A_4–A_5. The 80% of the molecules that couple successfully will yield the desired product. 20% of the molecules will lack A_3, but will couple in the subsequent steps to give the deletion peptide A_1–A_2–A_4–A_5.

2.4 Protecting Groups

An enormous amount of effort has been expended over the last 70 years in developing protecting groups for peptide synthesis, and it is not the intention to catalogue all of the reagents that have been used. A comprehensive review of earlier work in the field is available.[2] Here we will concentrate on some of those protecting groups that are most widely used in modern peptide synthesis, and in particular those used in solid-phase methods.

2.4.1 Amino Group Protection

The most important single advance in peptide synthesis was made in 1932 by Bergmann and Zervas,[3] who introduced the **benzyloxycarbonyl** protecting group. The essential point about benzyloxycarbonyl amino acids (or Cbz amino acids as they are commonly called from the – incorrect – designation of the protecting group as carbobenzoxy) is that, as esters of carbamic acid, the nitrogen atom does not have nucleophilic properties and will not take part in peptide bond formation. They are easily prepared by reaction with benzyloxycarbonyl chloride, as shown in Scheme 2.6. Equally important, the protecting group is removable

Scheme 2.6

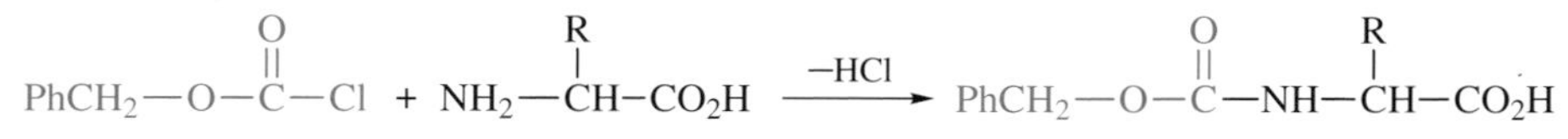

$$PhCH_2-O-\overset{O}{\overset{\|}{C}}-Cl \ + \ NH_2-\overset{R}{\overset{|}{C}H}-CO_2H \ \xrightarrow{-HCl} \ PhCH_2-O-\overset{O}{\overset{\|}{C}}-NH-\overset{R}{\overset{|}{C}H}-CO_2H$$

under a variety of conditions including treatment with liquid HF under conditions where peptide bonds are not affected.

Subsequently an improved carbamate protecting group, *t*-butyloxy-carbonyl (*t*-Boc), was developed. The structure of a *t*-Boc amino acid is shown in **4**. This group is removable by treatment with aqueous trifluoroacetic acid (TFA, trifluoroethanoic) under conditions milder than those required for Cbz. Availability of these two protecting groups removable under different conditions provides a strategy for synthesis of peptides containing lysine. The α-amino group can be protected with *t*-Boc and the side chain amino group with Cbz. During synthesis the α-amino group may then be deprotected with TFA for chain elongation under conditions where the side-chain amino protecting group is unaffected.

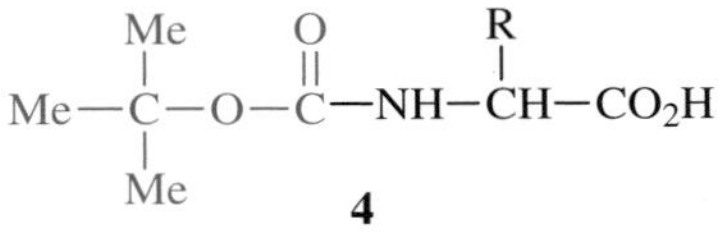

$$\underset{\underset{\overset{|}{Me}}{\overset{\overset{Me}{|}}{Me-C-O-}}}{} \overset{O}{\overset{||}{C}} -NH-\overset{\overset{R}{|}}{CH}-CO_2H$$

4

Worked Problem 2.3

Q Give a scheme for the reaction of N^α-*t*-Boc,N^ε-Cbz-lysine with TFA, showing the structure of reactants and products in full (note: N^ε is used to designate the side-chain nitrogen of lysine; the structure of lysine is given in Table 1.1).

A The *t*-Boc group is removed by treatment with TFA whereas the Cbz group is not. Removal of the *t*-Boc group occurs by an acid-catalysed elimination reaction, forming 2-methylprop-2-ene and carbon dioxide. Hence the reaction is as shown in Scheme 2.7:

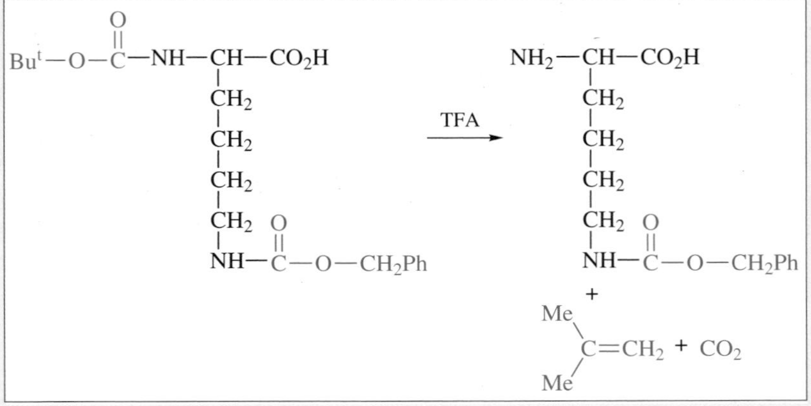

Scheme 2.7

A further important advance was made by Carpino and his co-workers,[4] who introduced the **9-fluorenylmethyloxycarbonyl** (Fmoc) protecting group **5**. The significance of this is that the protecting group is removable under very mild conditions by treatment with an organic base. This is because of the acidity of the proton at position 9 of the fluorenyl group. The deprotecting reaction is shown in Scheme 2.8.

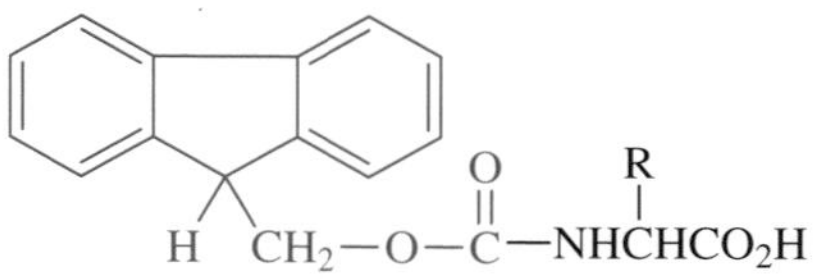

5

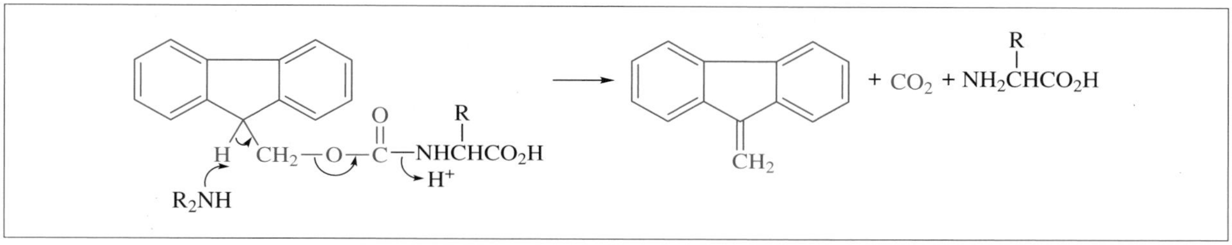

Scheme 2.8

2.4.2 Carboxyl Group Protection

Carboxyl group protection is usually carried out by conversion to esters. These must be unreactive in the process of peptide bond formation but easily removable during chain elongation or formation of the final deprotected product. Two widely used derivatives are **benzyl esters** (**6**) and *t*-butyl esters (**7**). The benzyl group can be removed by HF whereas the *t*-butyl group can be removed by treatment with TFA. Again it is important to have two protecting groups removable under different conditions because of the need for differential protection of amino acids with carboxylic acid side chains (aspartic acid and glutamic acid; see Table 1.1).

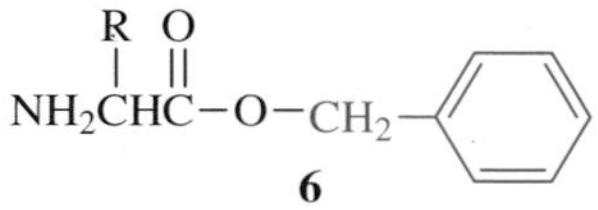

In the solid-phase method, of course, protection of the α-carboxylic acid function of the C-terminal residue is effected by the attachment to the solid support. The types of linkage used are dealt with in Section 2.6.

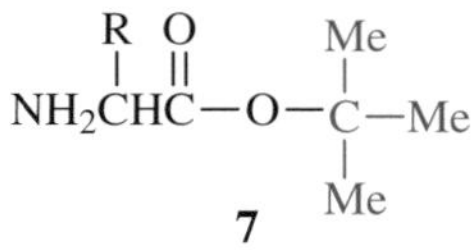

6

7

2.4.3 Side-chain Protection

Apart from those residues already mentioned (aspartic and glutamic acids, lysine) the side chains that are always protected are those of cysteine, serine, threonine, tyrosine, histidine and arginine. Whether or not tryptophan, methionine and the amino acids with amide side chains (asparagine and glutamine) are protected depends on the approach to peptide synthesis being used. We will concern ourselves here with the principles of side-chain protection rather than the details of the derivatives used in each case, but the interested reader can find further information in the articles referred to in the reading list at the end of the chapter.

The choice of side chain protecting agents will depend on the protecting group used for the α-amino function, since it is essential that the side chains remain protected when the N-terminal amino acid is deprotected for chain elongation. The two main approaches for SPPS are summarized in Table 2.1. When the *t*-Boc group is used for N^α-protection, then side chains are protected with groups based on the benzyl (phenylmethyl) function (*e.g.* the benzyl ether for threonine). Selective removal of the *t*-Boc group is achieved using TFA, which does not affect the side-chain protecting groups. The latter are removed with liquid HF. When Fmoc is used for N^α-protection, the side chains are protected with groups based on the *t*-butyl function (*e.g.* the *t*-butyl thioether of cysteine). Fmoc is removed by treatment with an organic base such as piperidine and the side chains are deprotected with HF.

Table 2.1 Summary of the two main approaches to side-chain protection in SPPS

N^α-protection	N^α-deprotection	Side-chain protection	Side-chain deprotection
t-Boc	TFA	Bz based	HF
Fmoc	Piperidine	*t*-Bu based	TFA

The *t*-Boc/benzyl strategy was the first to be developed and is very widely used. It does, however, have the disadvantage that the N^α-deprotection and side-chain deprotection are both carried out under acidic conditions, and there is the danger that the deprotection process may not be entirely specific. With the Fmoc/*t*-butyl strategy the two phases of deprotection are carried out under completely different conditions, and so specificity is more certain. It is also easier to automate. Both strategies have their adherents, and both are likely to continue in use for the foreseeable future.

Worked Problem 2.4

Q Draw the structures of the N-protected, side-chain-protected derivates of the amino acid serine that would be used in SPPS by (a) the *t*-Boc/benzyl strategy and (b) the Fmoc/*t*-butyl strategy.

A For the *t*-Boc/benzyl strategy the appropriate derivative of serine is the benzyl ether **8** whilst for the Fmoc/*t*-butyl strategy it is the *t*-butyl ether **9**.

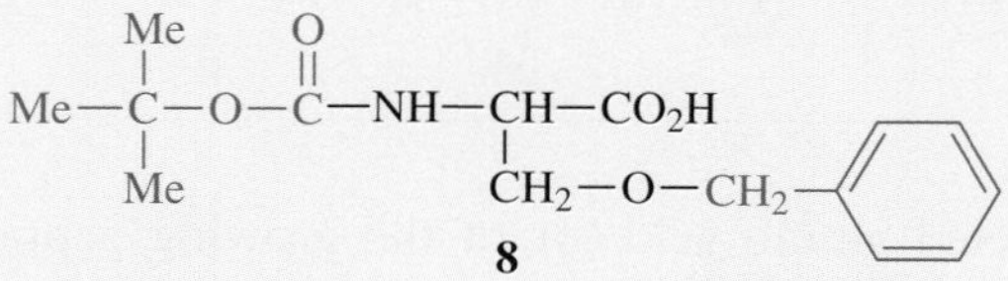

8

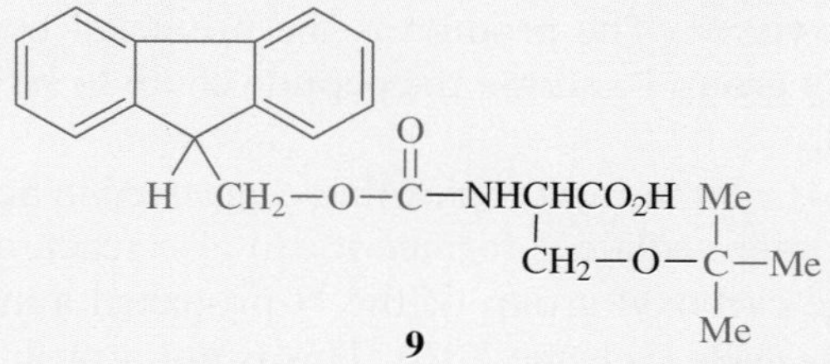

9

2.5 The Coupling Reaction

As outlined in Section 2.2, the basis of peptide bond formation is the conversion of the carboxylic acid function of one amino acid to a reactive acyl derivative that is susceptible to nucleophilic attack by the amino group of the second amino acid. The majority of methods currently in use are based on activation of the amino acid using **dicyclohexylcarbodiimide** (DCC, **10**).

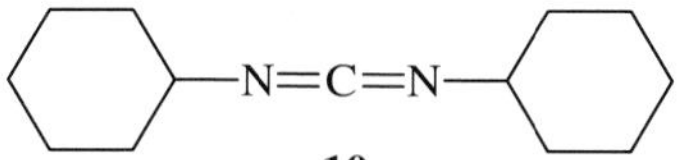

10

In the most straightforward application of DCC coupling, the protected amino acids are mixed in the presence of the coupling agent (Scheme 2.9; the cyclohexyl groups are shown as R).

The carboxylic acid function of the N-protected amino acid reacts with the DCC to form an *O*-acylisourea intermediate; this is effectively an **active ester** of the amino acid. In the second reaction the amino group

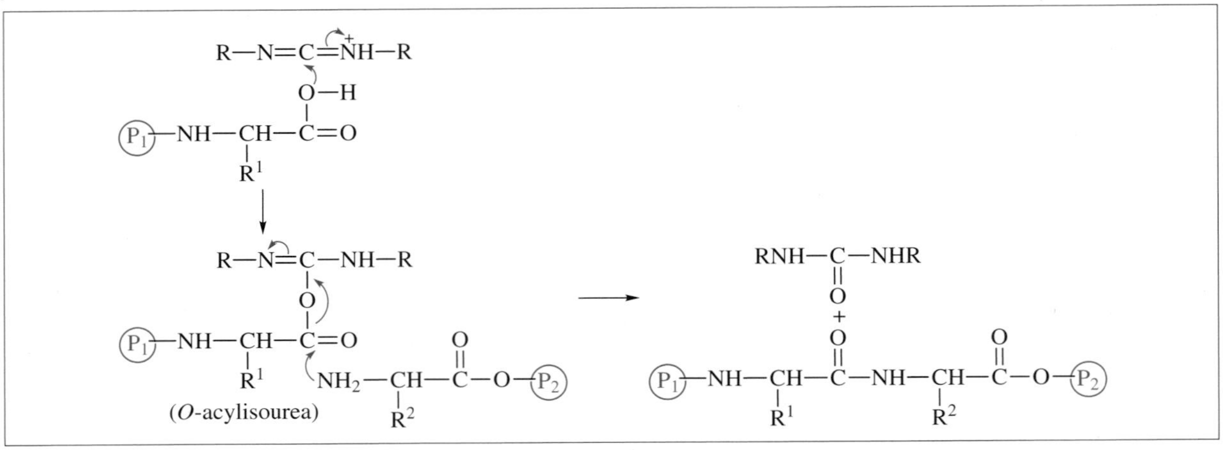

Scheme 2.9

of the C-protected amino acid (or of the growing peptide) attacks the carbonyl function of the intermediate to form a peptide bond and liberate dicyclohexylurea. The product is the protected peptide. Removal of the protecting group P_1 allows the peptide chain to be extended at the N-terminal end.

    The reaction is a bit more complex than suggested in Scheme 2.9. Once the O-acylurea intermediate is formed it can also react, albeit at a slower rate, with the carboxyl group of the N-protected amino acid to give a **symmetric anhydride** (Scheme 2.10). This is not a problem because the anhydride will then react with the C-protected amino acid to give the desired product. What can be a problem is the rearrangement of the O-acylisourea intermediate to give the unreactive N-acylurea (**11**), but this can minimized by appropriate choice of solvent.

Scheme 2.10

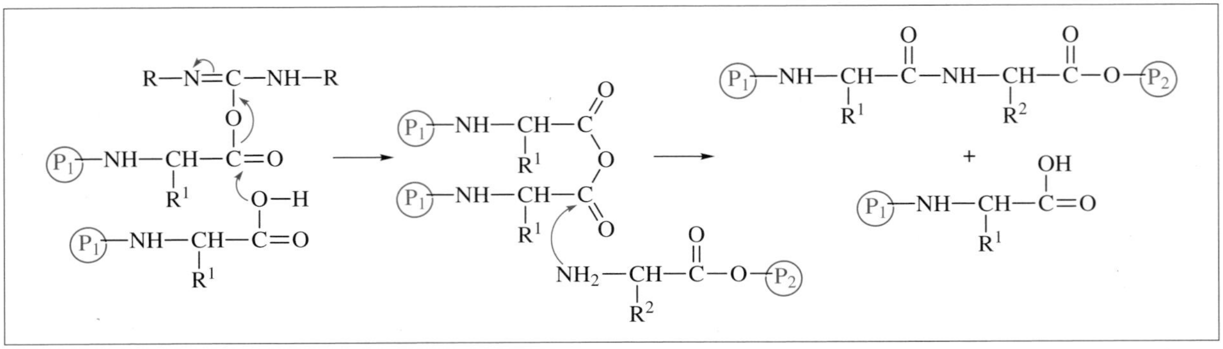

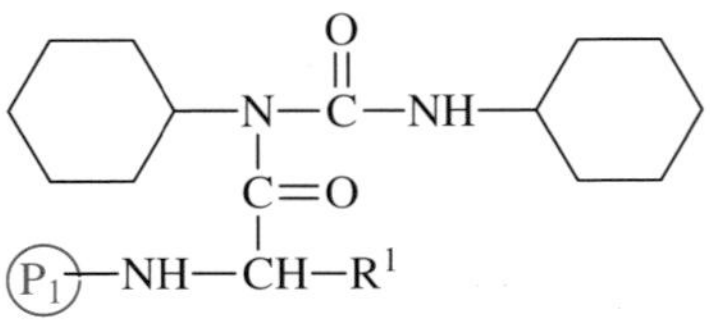

11

The formation of the symmetric anhydride by reaction of 1 mol of DCC with 2 mol of protected amino acid provides an alternative approach to peptide synthesis in which the anhydride is generated in a separate reaction and then mixed with the C-protected amino acid to generate the peptide. This method gives very high coupling yields but has the disadvantage that it requires the use of two molar equivalents of the N-protected amino acid at each step.

A more recent method of peptide bond formation, and one which is very widely used in automated SPPS, involves the DCC-mediated generation of an active ester between the N-protected amino acid and **1-hydroxybenzotriazole** (HOBT, **12**).[5] The pre-formed active ester is then reacted with the free amino group of the growing peptide attached to the solid support, as shown in Scheme 2.11.

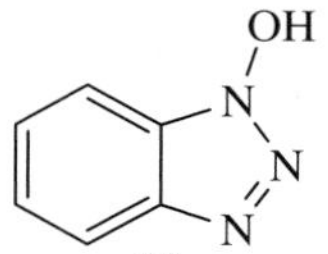

12

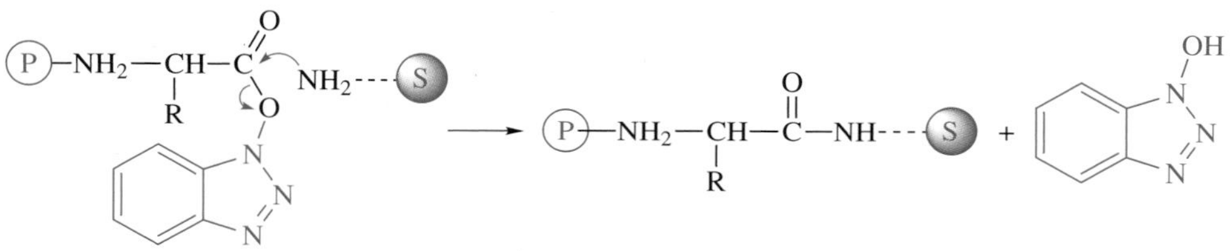

Scheme 2.11

Worked Problem 2.5

Q The 4-nitrophenyl esters of amino acids are reactive because of the electron withdrawing effect of the nitro groups. Consequently, they are sometimes used in the coupling step of peptide synthesis. Draw the structure of an N^α-t-Boc amino acid 4-nitrophenyl ester and show how it makes a peptide bond in peptide synthesis.

A The structure of the glutamine derivative and its reaction are shown in Scheme 2.14 in Section 2.9.

2.6 Linkage to the Solid Support in SPPS

In the original method developed by Merrifield, chloromethyl groups were introduced into the polystyrene resin and the C-terminal amino acid was attached by reaction with a trialkylamine salt to form an ester linkage (Scheme 2.12; note that benzene ring attached to the ball symbol is part of the polystyrene). Although some remarkable successes were achieved with this attachment method, it was not ideal because of the relatively harsh conditions required to attach the C-terminal amino acid in the first place and then to liberate the completed peptide at the end of the synthetic process. Consequently, much effort has been expended to the development of improved linkage chemistries.

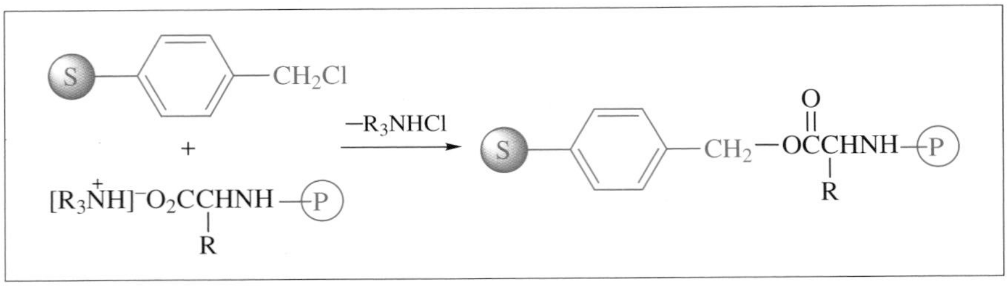

Scheme 2.12

One of the most widely used linkers, **4-alkoxybenzyl alcohol** (**13**), was developed by Wang[6] and resins with this group attached are now usually known as "**Wang resins**". Attachment of the C-terminal N-protected amino acid is easily effected by DCC coupling, and at the end of synthesis the product peptide can be released by treatment with aqueous TFA. Wang resin is ideally suited to Fmoc synthesis, since the linkage to the resin is not affected by the basic conditions required to release the 9-fluorenylmethoxy protecting group, whereas it is sensitive to acid. Indeed, a range of Wang resins is available commercially, each with a different Fmoc-protected amino acid already attached. A variant of the linker (**4-alkoxybenzyloxycarbonyl hydrazide**, **14**) is available. In this case the C-terminal amino acid is linked to the resin *via* an amide bond and on release by TFA generates the peptide amide. This is useful if the object of the synthesis is to produce a peptide amidated at the C-terminus.

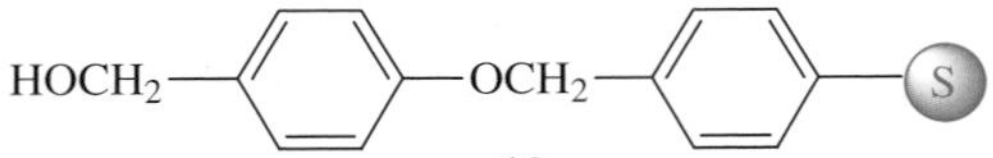

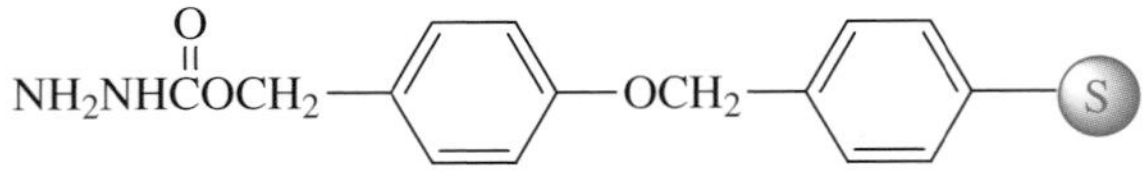

In SPPS using *t*-butoxy protection, the resin–peptide linkage must be stable to the acidic conditions used to deprotect the N-terminal amino acid. A suitable choice in this case is the **phenylacetamidomethyl** resin (PAM resin) shown in **15**. The final product of the synthesis is released by treatment with HF. The **benzhydrylamine** resin **16** is used to yield products with a C-terminal amide.

15

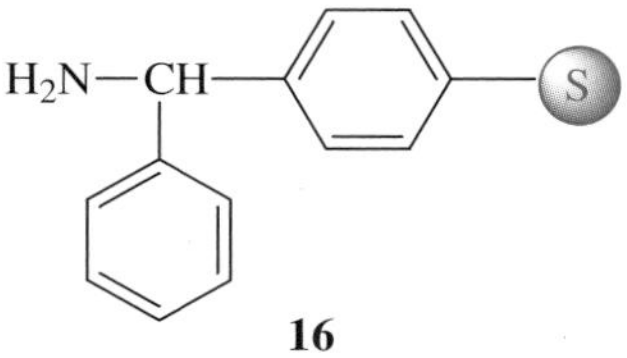

16

Worked Problem 2.6

Q Draw the structure of an N^α-t-Boc amino acid linked to benz-hydrylamine resin.

A The structure of N^α-t-Boc,N^{Im}-tosylhistidine attached to the benz-hydrylamine resin is given in Scheme 2.14 in Section 2.9.

2.7 Purifying and Characterizing the Product

After deprotection and release of the product peptide from the solid support it is invariably necessary to carry out some form of purification of the product peptide. The reasons for this are, firstly, that no matter how carefully the synthesis has been done to avoid deletion molecules arising from incomplete coupling (see Section 2.3), there are bound to be small amounts of such products present. Secondly, the deprotection process leads to small amounts of side products arising from modification of amino acid side chains. The usual way to purify the peptide is by preparative reverse-phase HPLC. Confirmation of purity of the product is usually done by using analytical HPLC with a range of different mobile phases.

Box 2.2 High Performance Liquid Chromatography

High performance liquid chromatography (HPLC)[7] is a separation technique capable of great resolving power. This is achieved by using very small particles for the solid phase so that the kinetics of exchange between the mobile and stationary phases is very rapid. The stationary phase particles have also to be regular in size to avoid the possibility of the material packing down and blocking the column. Even so, the small particle size means that the pressure required to pass liquid through the column is considerable and high-pressure pumps are required. In principle, any form of chromatography can be used, but in peptide chemistry by far the most important is **reverse-phase partition chromatography (RP-HPLC)**. The solid phase consists of particles coated with a hydrocarbon which may be 4, 8, 12 or 18 carbon atoms long. The peptide mixture is usually applied to the column in dilute aqueous TFA (to suppress ionization of carboxylic acid groups and to provide a counter ion for the basic groups), so that the peptides absorb onto the stationary phases by hydrophobic interactions (see Section 1.3.3). Elution is carried out by application of a gradient of increasing concentration of an organic solvent, frequently acetonitrile (ethanenitrile, MeCN). Peptides elute in order of their hydrophobicities.

As an example, Figure 2.2A shows RP-HPLC of the crude product of synthesis of a 49-residue peptide.[8] The effluent was monitored by absorbance at 220 nm. The desired peptide eluted at 14.96 min. The profile of the purified peptide is shown in part B. The elution time is earlier because a steeper gradient of acetonitrile was used.

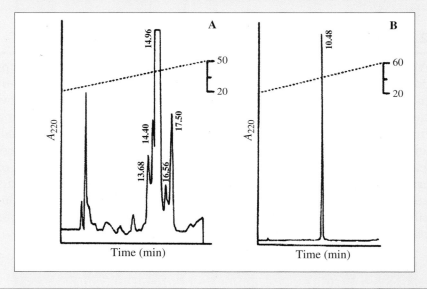

Figure 2.2 Reverse-phase HPLC of a crude product from peptide synthesis (A) and of the purified peptide (B). The scale on the right of each figure gives the percentage of acetonitrile in the gradient (Redrawn from Pennington[8] with the permission of Humana Press)

It is also good practice to prove that the peptide has the desired structure. A good indication of this would be to confirm that the product has the correct amino acid composition; how this is done is described in Chapter 4. A more rigorous test would be to determine the amino acid sequence; again, methods for doing this are given in Chapter 4.

2.8 Racemization During Peptide Synthesis

One of the most serious problems encountered in early peptide synthesis was partial **racemization** of the incorporated amino acids. Racemization is not easy to detect and, more importantly, it is not possible to separate peptides containing D-amino acids from the desired all-L molecules.

The main route to racemization is by **azlactone** formation from the carboxyl-activated amino acid derivative or peptide. This is shown in Scheme 2.13. Once the azlactone is formed it can lose a proton to form a resonance-stabilized anion which is achiral. Re-protonation of the anion can give either the azlactone of the original chirality or the enantiomer. Clearly this process will be helped by basic conditions, which therefore must be avoided.

The reason why racemization is such a serious problem is that biological activity is critically dependent on stereochemistry. It is likely that a synthetic peptide that contained even one D-amino acid would not be biologically active. If a significant fraction of each residue in a synthetic peptide were of the D configuration, the final product might well contain no all-L molecules and hence be completely inactive.

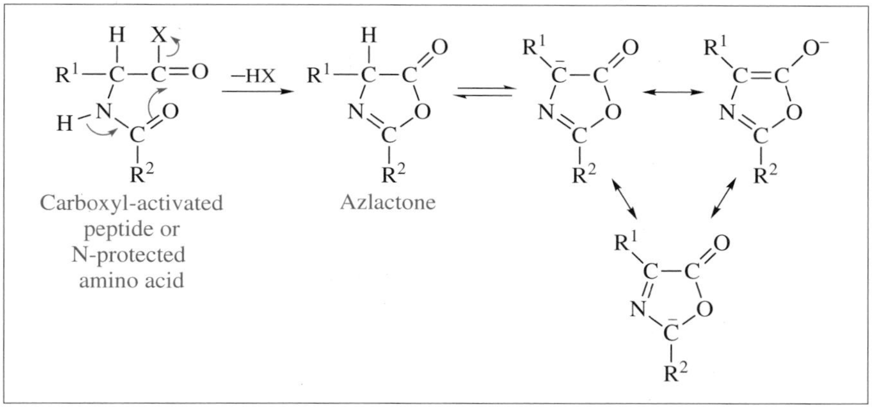

Carboxyl-activated peptide or N-protected amino acid

Azlactone

Scheme 2.13

Fortunately this process does not occur to any significant extent with amino acids N-protected by *t*-Boc or Fmoc groups. Similarly, it does not occur during elongation of a peptide from the C-terminal end, because the carboxyl function is protected by an unreactive group, or during SPPS because the carboxyl group of the C-terminal amino acid is bound to the solid support in an unreactive linkage. On the other hand, synthesis by chain elongation from the N-terminal end [Scheme 2.2, reactions (v) and (vi)] involves activation of the carboxylic acid function at each step, and there is always a risk of racemization. Hence this approach to synthesis is rarely used.

Even though the risks of racemization are minimized by synthesis of peptides from the C-terminus and with the appropriate chemistry, there remains a problem when large peptides are made by the fragment condensation method (Scheme 2.3). Once the fully protected fragments have been produced, either by solution synthesis or SPPS, they have to be coupled and this inevitably involves activation of the carboxylic acid group of all but the C-terminal fragment. There is, then, the attendant risk of racemization. In favourable circumstances this can be avoided by making fragments with C-terminal glycine or proline residues. Glycine is not stereogenic and so the problem does not arise, and proline, being an imino acid, lacks the proton that is removed in azlactone formation. If neither of these is possible, then it is necessary to try a variety of coupling methods for the fragments and determine the one that gives the minimum amount of racemization for the fragments under consideration.

Worked Problem 2.7

Q Draw the structure of a carboxyl-activated, N-protected derivative of proline and explain why it cannot form an azlactone.

A The structure is shown as **17**. Remember that proline is an imino acid and that the imino nitrogen and the α-carbon atom are part of a five-membered ring. The derivative lacks a hydrogen at the position indicated by the arrow and hence, by analogy with Scheme 2.13, cannot form an azlactone.

$$
\begin{array}{c}
\text{H}_2\text{C}\overset{\text{CH}_2}{\diagdown}\quad\quad\overset{\text{X}}{|} \\
|\quad\quad\text{CH}-\text{C}=\text{O} \\
\text{H}_2\text{C}\diagdown_{\text{N}}\diagup\leftarrow\text{— No removable H} \\
\text{C}=\text{O} \\
\text{R}\quad\quad\textbf{17}
\end{array}
$$

2.9 An Example of Peptide Synthesis

Peptide synthesis is obviously a complicated process, and it might be helpful to give an example of how it was done in practice. The peptide apamin (**18**) is a neurotoxin that occurs in high concentration in the venom of the honey bee. Notice that it has two disulfide bridges linking residues $1\rightarrow11$ and $3\rightarrow15$ and that the C-terminus is amidated The peptide was synthesized by van Rietschoten and colleagues[9] using SPPS by the procedure summarized in Table 2.2. The *t*-Boc/benzyl strategy was used and the linker, benzhydrylamine, was dictated by the fact that the

$$
\begin{array}{c}
\overline{\quad\quad\text{S}\quad\quad\text{S}\quad\quad} \\
\text{Cys-Asn-Cys-Lys-Ala-Pro-Glu-Thr-Ala-Leu-Cys-Ala-Arg-Arg-Cys-Gln-Gln-HisNH}_2 \\
\underline{\quad\quad\quad\quad\text{S}\quad\quad\text{S}\quad\quad\quad\quad} \\
\textbf{18}
\end{array}
$$

C-terminus of the peptide is amidated. Histidine and arginine were protected as the tosyl (4-toluenesulfonyl or 4-methylbenzenesulfonyl) derivatives; the structures of the protected side chains are shown in **19** and **20**. Coupling was by the DCC method (Scheme 2.9) except for the glu-

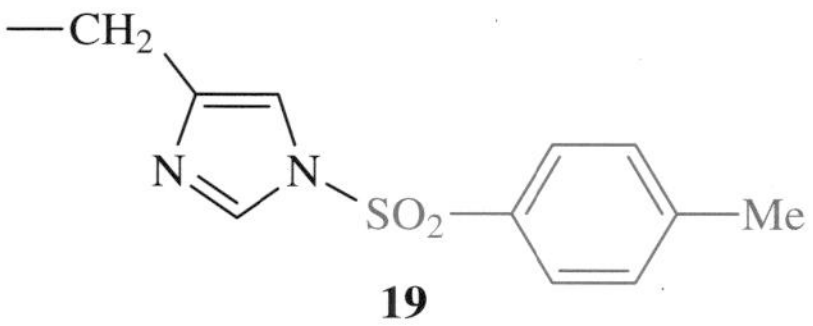

19

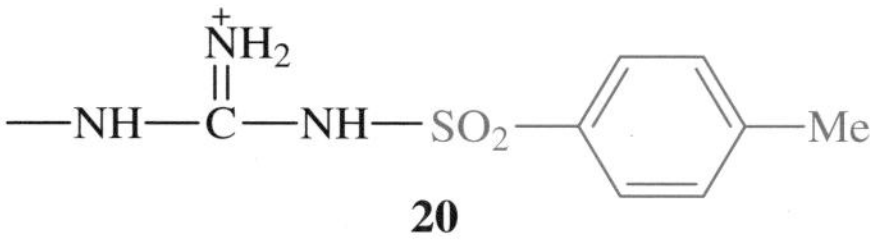

20

tamine and asparagine, where 4-nitrophenyl esters were used. The first two cycles of synthesis are shown in Scheme 2.14. Double coupling was carried out at each step and the completeness of coupling checked by testing for free NH_2 groups. Coupling at two positions (Cys-11 and Gln-17) did not go to completion and residual amino groups were blocked by treatment with acetylimidazole (Scheme 2.15). Final release of the peptide from the resin and deprotection were done by treatment with HF. The amino acid sequence of the product was determined by Edman degradation (see Chapter 4), and shown to be identical with that of the natural molecule. The average yield of incorporation of residues at each step was 99.3%.

Apamin presents a problem that we have not deal with previously; that is, it contains disulfide bridges (see Section 1.4) and these have to

Table 2.2 Experimental conditions for the synthesis of apamin

Step	Method
N^α-protection	t-Boc
N^α-deprotection	30% TFA in CH_2Cl_2
Linker	Benzhydrylamine
Coupling	DCC
Side-chain protection	Thr: benzyl ether
	Glu: benzyl ester
	Lys: benzyloxycarbonyl
	Cys: t-butyl thioether
	His: tosyl
	Arg: tosyl
Removal from resin and deprotection	HF, 1 h at 0 °C

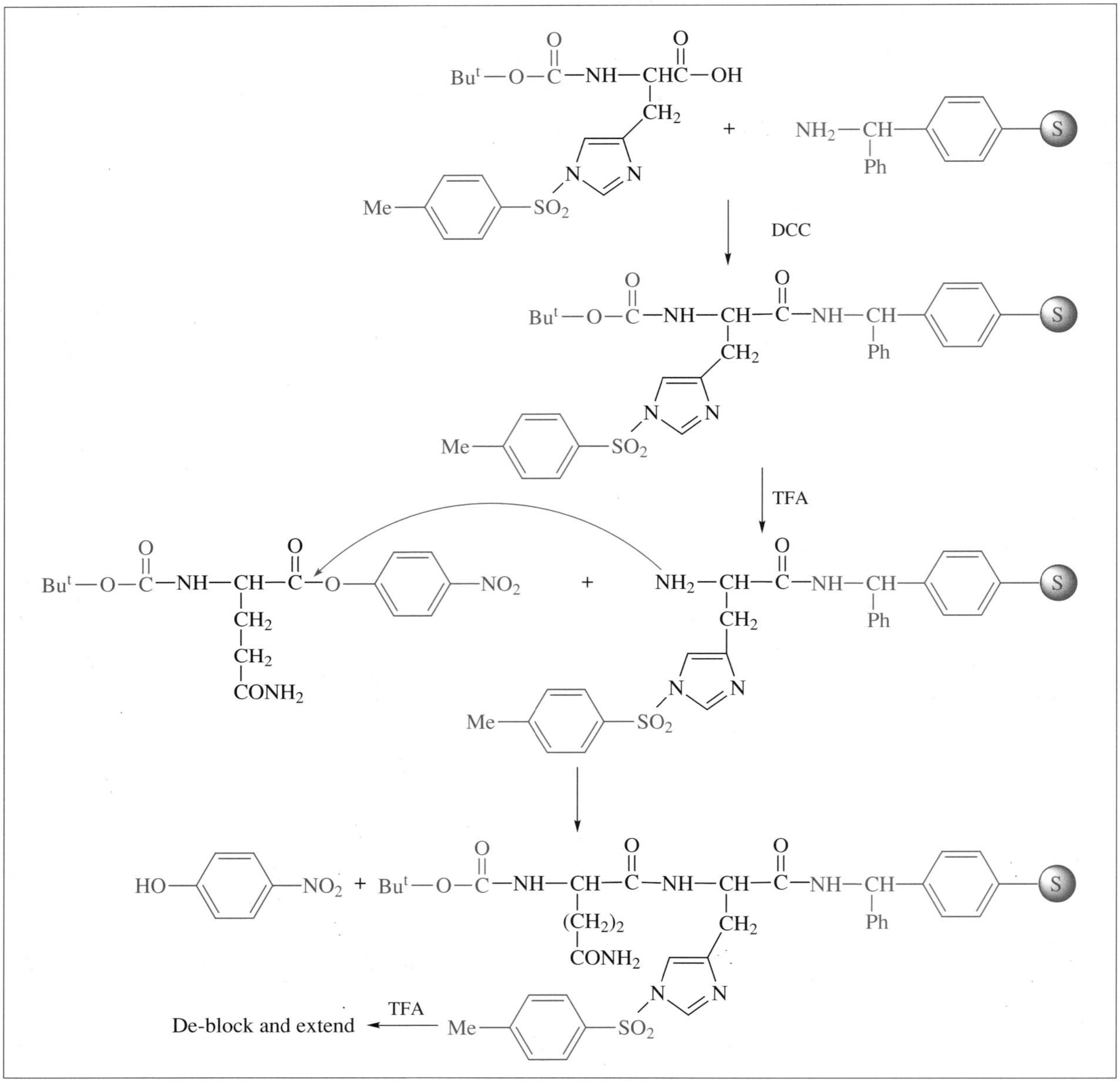

Scheme 2.14

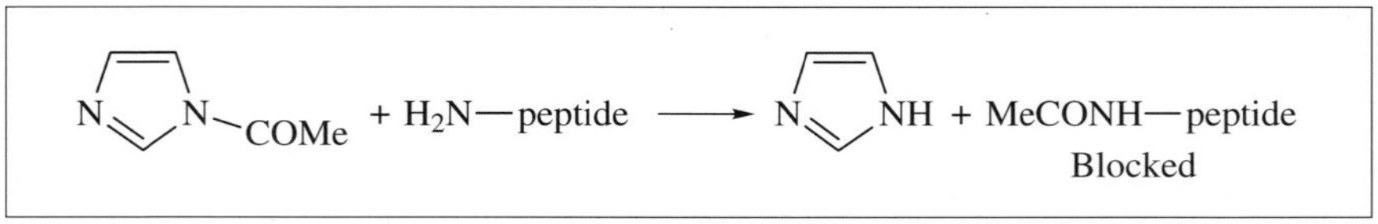

Scheme 2.15

be formed by oxidation of the incorporated cysteines. The product peptide was first fully reduced by treatment with the reducing agent tributylphosphine (some degree of oxidation of cysteines during synthesis and deprotection usually occurs). After removal of the reducing agent the peptide in dilute solution (to avoid intermolecular disulfide formation) was oxidized with air for 48 h to form the disulfide bridges. Note that in principle there are three ways that the bridges could form ($1 \rightarrow 11$ and $3 \rightarrow 15$; $1 \rightarrow 3$ and $11 \rightarrow 15$; $1 \rightarrow 15$ and $3 \rightarrow 11$). In fact, only the correct pairing was produced, which presumably means that the reduced peptide adopts the native three-dimensional shape (see Chapter 5), which brings the cysteines into the correct positions to form the bridges.

Summary of Key Points

- Chemical synthesis of peptides and their structural analogues is important as an aid to understanding how they carry out their biological functions, and also so that they can be made available for therapeutic use.
- Central to the process of peptide synthesis is the use of protecting groups that prevent reaction of those functional groups of the amino acid not intended to participate in a particular part of the process. The protecting groups must be selectively removable.
- Of particular importance in the synthesis of large peptides is that the process of peptide bond formation at each stage must be essentially complete, otherwise the final yield will be unacceptably low.
- In solid-phase peptide synthesis the peptide chain is built up with the C-terminal amino acid linked to a solid support. SPPS requires much less chemical skill than does solution synthesis and can be automated. It is the method of choice for all but specialist peptide chemists.
- The choice of linkage to the solid support and of side-chain protecting groups in SPPS is dictated by the N^{α}-protecting group being used.
- Most methods of peptide bond formation involve the use of dicyclohexylcarbodiimide as a coupling agent to form active esters or anhydrides.
- Purification of synthetic peptides is usually carried out by RP-HPLC.
- Recemization is potentially a severe problem in the production of large peptides by block synthesis and fragment condensation. It can be overcome by synthesizing blocks of sequence with glycine or proline as the C-terminal residue.

Problems

2.1. Referring to Scheme 2.2, give the structure of the protected and activated amino acid that would be required in reaction (iv), and show the structure of the product of the reaction.

2.2. In Worked Problem 2.1 it was shown that, in the synthesis of a 25-residue peptide with a step yield of 95%, the final yield would be 29.2%. Calculate the overall yield if the peptide was synthesized by the fragment condensation method in two blocks of 12 and 13 residues and if all step yields, including fragment condensation, were 95%.

2.3. Referring to Scheme 2.5 and starting from the last structure shown, complete the scheme for the synthesis of a tripeptide (three amino acids).

2.4. Give a scheme for the reaction of N^α-Fmoc,N^ε-t-Boc-lysine with piperidine (azacyclohexane, **21**), showing the structures of reactants and products in full.

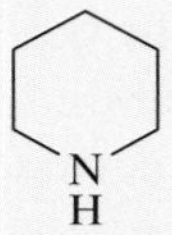

N
H
21

2.5. Draw the structures of the N-protected, side-chain-protected derivatives of aspartic acid that would be used in SPPS by (a) the t-Boc/benzyl strategy and (b) the Fmoc/t-butyl strategy.

2.6. The 4-nitrophenyl esters of N-protected amino acids can be used in peptide synthesis. Give a reaction sequence for their synthesis from the N-protected amino acid and 4-nitrophenol by DCC coupling.

2.7. Suggest a reaction scheme for the attachment of Fmoc glycine to Wang resin.

2.8. Starting with glycine attached to Wang resin, outline the steps required to synthesisze the peptide Ala-Glu-Gly using HOBT coupling.

References

1. R. B. Merrifield, *J. Am. Chem. Soc.*, 1963, **85**, 2149.
2. H.-D. Jakubke and H. Jeschkeit, *Amino Acids, Peptides and Proteins*, Macmillan, London, 1977, p. 37.
3. M. Bergmann and L. Zervas, *Ber. Dtsch. Chem. Ges.*, 1932, **65**, 1192.
4. L. A. Carpino and G. Y. Han, *J. Am. Chem. Soc.*, 1970, **92**, 5748.
5. W. König and R. Geiger, *Chem. Ber.*, 1970, **103**, 788.
6. S.-S. Wang, *J. Am. Chem. Soc.*, 1973, **95**, 1328
7. D. Sheehan, *Physical Biochemistry*, Wiley, Chichester, 2000, p. 46.
8. M. W. Pennington, in *Peptide Synthesis Protocols*, ed. M. W. Pennington and B. M. Dunn, Humana, Totowa, NJ, 1994, p. 41.
9. J. van Rietschoten, C. Granier, H. Rochat, S. Lissitzky and F. Miranda, *Eur. J. Biochem.*, 1975, **56**, 35.

Further Reading

An on-line tutorial on peptide synthesis, written by Dr Mike Porter of University College London, is available at http://teach.chem.ucl.ac.uk/mjp/peptide/simulation.html. If you visit the site, be prepared for some differences in the chemistry used compared with here.

F. M. Finn and K. Hofman, in *The Proteins*, ed. H. Neurath and R. L. Hill, 3rd edn., Academic Press, New York, 1976, pp. 105–253.

B. W. Erickson and R. B. Merrifield, in *The Proteins*, ed. H. Neurath and R. L. Hill, 3rd edn., Academic Press, New York, 1976, p. 255.

M. Bodansky, *Peptide Chemistry*, 2nd edn., Springer, Berlin, 1993.

3
Protein Purification

Aims

By the end of this chapter you should understand:

- Why the purification of proteins presents unique problems
- That purification of proteins depends on exploiting the ways in which proteins differ from one another
- The purification techniques that arise from each of these differences and how they are carried out
- How the various techniques can be combined into a practical schedule for protein purification

3.1   Nature of the Problem

The problems involved in the purification of proteins are rather different from those experienced by the synthetic chemist. At any step during an organic synthesis the desired intermediate, or the final product, should constitute at least the majority of the sample, and the impurities will generally be compounds of a different chemical type. Purification can then be achieved using the standard methods of distillation, crystallization, and so on.

The protein chemist faces a rather more difficult problem. A typical cell contains hundreds or thousands of different proteins. This means that, if we make an extract of a tissue, the protein that we are interested in may be present at a level of only 0.1% or even less of the total protein in the extract. Moreover, all of the other components of the mixture are of the same chemical type: they are all proteins. So the problem is how to separate a particular protein present at a very low level in a mixture where all the contaminants are also proteins.

There is another problem. Proteins are delicate molecules. The main

The complete structure of the human genome has recently been determined. This work has shown that humans make at least 30,000 different proteins, although not all of these proteins will be expressed in any particular tissue.

reason for this is that their biological activity depends on the way in which they fold up in space to form precisely defined three-dimensional structures (see Chapter 5). These three-dimensional structures are very easily disrupted – a process known as **denaturation** – and so proteins can only be handled under a restricted range of conditions. This usually means at or below room temperature, in aqueous solutions, with a pH not far removed from neutrality.

Box 3.1 Buffers

Buffers are solutions that resist changes in pH when acid or base is added. Their use is essential in the vast majority of experiments involving proteins, and particularly those involved with protein purification. Buffers are solutions of a weak acid and its salt, or a weak base and its salt. Consider a solution of acetic (ethanoic) acid and sodium acetate (ethanoate) as an example. In such a solution the acetic acid will be dissociated to only a very small extent (Scheme 3.1), whereas the sodium acetate will be essentially completely dissociated (Scheme 3.2). If H_3O^+ is added to the buffer then the equilibrium in Scheme 3.1 moves to the left, the acetate ions being provided by the dissociated sodium acetate. If, on the other hand, OH^- is added, then the base combines with H_3O^+ provided by further dissociation of the acetic acid to form water. In either case the effect on the pH of the solution is minimized.

$$MeCO_2H + H_2O \rightleftharpoons MeCO_2^- + H_3O^+$$

Scheme 3.1

$$MeCO_2Na \longrightarrow MeCO_2^- + Na^+$$

Scheme 3.2

It is easy to calculate the pH of a buffer made from a weak acid and its salt by reference to equation (1.4) in Chapter 1. The term $[A^-]$ will be very nearly equal to the concentration of salt in the buffer because the salt is essentially completely ionized. The term $[HA]$ will be very nearly equal to the concentration of acid in the buffer because the acid is only slightly dissociated. Hence the pH is given by equation (3.1):

$$pH = pK_a + \log\frac{[salt]}{[acid]} \qquad (3.1)$$

This is one form of what is known as the **Henderson–Hasselbalch** equation. A more general form is equation (3.2):

$$pH = pK_a + \log \frac{[\text{proton acceptor}]}{[\text{proton donor}]} \qquad (3.2)$$

Worked Problem 3.1

Q What is the pH of a buffer made by mixing 500 mL of 0.200 mol dm^{-3} acetic acid and 500 dm^3 of 0.100 mol dm^{-3} sodium acetate? Take the pK$_a$ of acetic acid as 4.76.

A The final concentrations of acetic acid and of sodium acetate are 0.100 mol dm^{-3} and 0.0500 mol dm^{-3}, respectively. Putting these values in the Henderson–Hasselbalch equation leads to:

$$pH = 4.76 + \log \frac{0.0500}{0.100} = 4.46$$

Worked Problem 3.2

Q How much sodium hydroxide must be added to 1.00 dm^3 of 0.100 M acetic acid to make a buffer of pH 5.00?

A Let the molarity of NaOH required be x mol dm^{-3}. The concentration of acetic acid remaining will be $(0.100 - x)$ mol dm^{-3} and the concentration of salt will be x mol dm^{-3}. Hence:

$$5.00 = 4.76 + \log \frac{x}{0.100 - x}$$

Solving this equation gives $x = 0.064$ mol dm^{-3}. Hence the amount of NaOH required is 0.064 mol.

Small peptides are much more robust because they do not have defined three-dimensional structures. They can, therefore, be handled by the techniques more familiar to the organic chemist. For example, extraction of a tissue with dilute acid will normally produce a solution containing the peptides in the tissue, but not the larger proteins which will have been denatured and rendered insoluble. The mixture will not be very complex, because tissues contain relatively small numbers of peptides, and the components can be separated by RP-HPLC as described for synthetic peptides (Section 2.7). Because there are no new problems involved in the purification of peptides, we will not deal with the topic further.

3.2   Approaches to a Solution

Given that the impurities in a protein extract are themselves all proteins and have the same basic structure, the solution to the problem of how to purify proteins must come from the question: how do proteins differ from one another? The question then becomes: how can we exploit those differences to achieve separation of one protein from another?

A summary of the ways in which proteins differ from one another, and the ways that these differences can be used in purification, is given in Table 3.1. Some explanation is needed before we deal with the details of the purification methods.

Table 3.1 Differences between proteins and their exploitation for purification

Property	Resulting differences	Purification method
Amino acid composition	Electrical charge	ion exchange chromatography
	Surface hydrophobicity	Hydrophobic chromatography
	Solubility in salt solutions	Fractional precipitation
Binding sites	Affinity for other molecules	Affinity chromatography
Carbohydrate content		Lectin affinity chromatography
Antigenic determinants		Immunoaffinity chromatography
Number of residues	Size	Gel permeation chromatography

A fundamental difference between proteins is that they have different amino acid compositions. Given that some of the amino acids are ionizable (see Section 1.3.2), this means that, at a particular pH, different proteins will have different net electrical charges. This fact is exploited in **ion exchange chromatography**, which is the most generally applicable and most important of all the techniques of protein purification. Another difference between amino acids is that some are hydrophobic and some hydrophilic (see Section 1.3.3). Although most of the hydrophobic amino acids are found in the interiors of proteins (see Chapter 5), some are exposed on the surface and lead to regions with a hydrophobic character. This allows the possibility of using **hydrophobic interaction** chromatography as a means of separation. This technique is similar in principle to the RP-HPLC method that we have already come across (see Chapter 2).

Box 3.2 Column Chromatography

The majority of methods in protein purification involve the use of **column chromatography**. A block diagram of the equipment required is shown in Figure 3.1. The column, usually of glass, contains an insoluble inert supporting medium termed the **matrix**. This is chemically modified with the appropriate groups for the type of chromatography to be carried out, and is equilibrated with a suitable buffer; specific examples are given in Section 3.3. The sample is applied to the column, and those components not absorbed are eluted from the column by pumping the starting buffer through it from the reservoir. The eluent is passed through a UV absorption detector which measures the protein content of the solution, and the detector output is fed to a chart recorder. This gives a continuous record of the protein content of the eluent. The eluent is finally directed to a fraction collector.

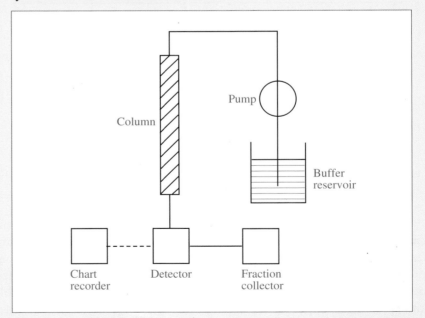

Figure 3.1 Block diagram of equipment used for column chromatography. The solid lines represent tubing through which liquid passes. The dashed line is an electrical connection

When no more protein is eluted with the starting buffer, the solution is changed to one that will elute the bound proteins. This may be, for example, a buffer of different pH, or the starting buffer containing a salt. Frequently, a solution of continuously increasing salt concentration is used; this is referred to as **gradient elution**. Elution is continued until all the protein has been eluted from the column.

The equipment used for column chromatography can be more sophisticated than that shown. For example, it is possible to connect the fraction collector to the detector so that fractions are only collected when protein is present in the eluent. Equally, it can be much simpler. In principle, all that is required is the column. All the rest can be done manually. The practicalities of column chromatography have recently been reviewed.[1]

Another characteristic of the vast majority of proteins is that they have specific binding sites for other molecules. This is a reflection of the way that proteins work. To exert their actions, proteins have to bind to the molecule on which they are going to act. For example, for an enzyme to catalyse its specific reaction, it must first bind the molecule (referred to as the **substrate**) on which it acts at a particular site on the enzyme (the **active site**). The active site of an enzyme will generally recognize and bind only one molecule, or a very small number of structurally related molecules. Similarly, any particular molecule will only be recognized by one enzyme, or a small number of different enzymes. To give some other examples, hormones often exert their biological actions by binding to specific **receptor proteins** in the membranes of their target cells. Proteins that are involved in regulation of expression of the genetic messages in cells do so by recognizing, and binding to, specific sequences of the chemical units of which the genes are made. Some proteins contain binding sites for metal ions; these may be involved in the mechanism of action of the protein or, in some cases, have a structural role. This property of specific recognition of other molecules is exploited in a set of powerful techniques collectively known as **affinity chromatography**.

Immunoaffinity chromatography is a special case of affinity chromatography. It relies on the extremely highly specific binding of an **antigen** to an **antibody**.

Box 3.3 Antibodies

The **immune system** is our primary defence against foreign agents such as bacteria and viruses. The invading agents display structures called **antigenic determinants** that are recognized by the immune system as non-self, and the system then sets about removing the foreign species. One component of the immune defence is provided by the **immunoglobulins**, or **antibodies**. These molecules are produced by specialized blood cells called **B cells**, each of which produces a different antibody molecule and displays it on the cell surface. If the antibody recognizes the antigenic determinant, then the B cell proliferates and secretes antibody molecules into the blood stream. There they bind to the invading agent and essentially label it to be destroyed by specialized cells called **macrophages**. Not only does the immune system remove the foreign agent, but it also "remembers" having done so, and if presented with the same agent on a subsequent occasion, is primed to mount a more rapid and much stronger defence against it. This is the basis of the processes of **immunization** against infectious diseases. The immune system is enormously complex and this brief account does no more than scratch the surface; interested readers can learn more from standard texts, such as that by Roitt.[2] We will look further at the structures of immunoglobulins and how they recognize foreign agents in Chapter 5.

For our present purposes the immune system provides a tool for producing highly specific reagents. If, for example, a particular human protein is injected into the blood stream of a rabbit, then the rabbit recognizes it as foreign, and produces antibodies against it. The rabbit does this even though the human protein will differ only slightly in structure from the same protein produced by the rabbit. If we now isolate the rabbit anti-human antibodies from the rabbit's blood, we will have a reagent which recognizes and binds very specifically to the human protein.

So far, we have considered proteins to consist solely of amino acids, and in many cases this is in fact correct. Some proteins, however, have non-amino acid components that are added after synthesis of the polypeptide chain. One example is provided by the class of molecules called **glycoproteins**, which have polysaccharide chains attached to some of the amino acids. These polysaccharide chains have very diverse structures and serve a range of functions, but it is beyond the scope of this book to deal with these matters in detail. For present purposes it is sufficient to note that some plants produce proteins called **lectins** that have the property of recognizing, and binding to, specific saccharides or polysaccharide chains. This property offers the possibility of yet a further type of affinity chromatography known as **lectin affinity chromatography**. This technique can be used to separate glycoproteins from those that do not contain carbohydrate, and also to separate individual glycoproteins, depending on the types of polysaccharide chains that they contain.

Perhaps the most obvious distinguishing feature between proteins is size: proteins vary in size from a few tens to a few thousands of amino acids. The associated separation method is **gel permeation chromatography** (also known as **size exclusion chromatography** or **gel filtration**). This method separates proteins on the basis of the effective radii of the spheres that they generate on rotation in solution, but for most proteins this is related to the number of amino acids that they contain. In spite of the very wide range of sizes that proteins show, gel permeation chromatography is not a very powerful separation method, for reasons given in Section 3.3.6.

3.3   Specific Techniques

3.3.1 Extraction

The first problem in protein purification is to obtain an aqueous extract containing the protein of interest. This is not difficult when the source is an animal tissue; in such cases it is sufficient to blend the tissue in a suitable buffer using a domestic food mixer and then to remove insoluble material, such as connective tissue, by **centrifugation**. In other cases, for example where the source is a bacterium, yeast or plant material, it is more difficult because these organisms have tough cell walls that are not easily disrupted by mechanical means. How these problems are solved, and indeed what determines the choice of source in any particular case, are matters beyond the scope of this book, but the methods have recently been described in detail (see Doonan[3] and other chapters in the same volume). The following sections describe the techniques that are used to separate the components of the protein extract.

Box 3.4   Centrifugation

A **centrifuge** is a device for spinning samples contained in (usually) plastic vessels, and held in a **rotor**, around a central axis. This develops a **centrifugal field** in which particulate matter in the sample migrates down through the liquid to the bottom of the tube. At the end of the centrifugation the liquid can be poured off and the **pellet** of precipitated material re-dissolved. A range of centrifuges is found in biochemistry laboratories, capable of handling volumes of solution from a few millilitres to a few litres. They will generally be refrigerated so that the sample can be kept cool during centrifugation.

The centrifugal force experienced by a particle in a centrifuge varies with the angular velocity of the rotor and the distance from the centre of rotation. If the angular velocity is ω revolutions per minute (rpm) and r is the distance from the centre of rotation in cm, then the **relative centrifugal field** (RCF, the centrifugal force compared with that of the Earth's gravitational field, g) is given by:

$$RCF = r\omega^2 \times 1.119 \times 10^{-5}\, g$$

Centrifugation conditions are always quoted in terms of the RCF and the time for which the centrifugation was carried out so that the experiment can be replicated using a different machine in a different laboratory.

3.3.2 Fractional Precipitation

This technique depends on the fact that different proteins precipitate out of solution at different salt concentrations. The salt usually used is ammonium sulfate because it is very soluble in water, and also because ammonium ions and sulfate ions are innocuous to most proteins.

It is conventional, when describing the conditions for salt fractionation, to express the concentration of ammonium sulfate in terms of **percentage saturation**. So, for example, the phrase "75% ammonium sulfate was added" means that 75% of the amount required to produce a saturated solution was added. Similarly, the phrase "35–50% ammonium sulfate fraction" means the protein that was precipitated between 35% saturation and 50% saturation.

The method is crude because precipitation for any particular protein occurs over a rather wide range of salt concentrations, and in complex mixtures the ranges will overlap. A hypothetical example is shown in Figure 3.2, where the precipitation curves are shown for a mixture containing only three proteins, one of which (B) we wish to purify. We might proceed as follows:

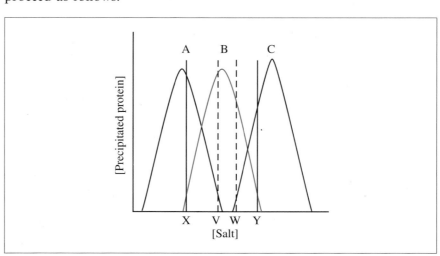

Figure 3.2 Precipitation curves for three proteins (A, B and C) as a function of salt concentration

(1) Add salt to a concentration represented by point X on the *x*-axis. Some of protein A will precipitate.

(2) Remove the precipitate from the suspension by centrifugation and collect the supernatant liquid.

(3) Add more salt to the supernatant to bring the concentration up to point Y. All of B will precipitate plus the remainder of A and some of C.

(4) Collect the precipitate by centrifugation. Pour off the supernatant and re-dissolve the precipitate.

This will clearly yield a product which is purer that the starting material in that a proportion of both A and C have been removed, but the **purification factor** will be rather small (see Section 3.4 for further discussion of what is meant by this term). We could proceed differently by adding salt to concentration V at the first step, which will precipitate nearly all of A but also some of B, and then increasing the concentration of salt to W, which will precipitate very little of C but will also leave some of B in solution. The result in this case is a purer product but at a substantially reduced **yield**. For reasons that will be returned to in Section 3.4, a high **step yield** is usually of considerable importance, and increased purity will normally be sacrificed in the interest of high yield.

It might seem that fractional precipitation is of limited value. It is, however, a very useful technique, particularly when working on a large scale. If, for example, 5 kg of tissue are used, about 10 dm^3 of buffer would be required for the initial extraction. This is too large a volume for chromatographic work using normal laboratory equipment – water is a major contaminant in the initial extract! However, the protein collected by centrifugation during salt fractionation can be re-dissolved in a much smaller volume of buffer – say 500 cm^3 – so achieving a 10-fold reduction in the volume of the solution for the next step as well as a modest purification.

In the early days of protein purification, methods depending on differential solubility of proteins in various solutions were very widely used. Some remarkable successes were achieved. For example, James Sumner purified the enzyme urease from jackbean meal by extracting the meal with aqueous acetone (propanone), from which the enzyme was obtained in crystalline form. He was awarded the Nobel Prize for Chemistry in 1946 for showing that enzymes could be crystallized.

Worked Problem 3.3

Q During salt fractionation the precipitated protein was collected by centrifugation at 5000 rpm in a rotor of radius 30 cm. What was the RCF?

A ω = 5000 rpm and r = 30 cm. Hence the RCF is given by 30 × 5000^2 × 1.119 × 10^{-5} = 8400 g.

3.3.3 Ion Exchange Chromatography

This is by far the most general and powerful method available for the purification of proteins. The solid supports used for protein work are usually polysaccharides such as cellulose, dextran or agarose, on to which the ion exchange groups are substituted. Examples of these groups are given in Table 3.2. The exchangers are usually classified as weak or strong, depending on the pH range over which they can be used. The carboxymethyl (CM) group has a pK_a value of about 4, and so is useful as an ion exchanger at pH values of 5 and above. For work at a lower pH, the strong cation exchanger propylsulfonate (SP), with a pK_a value of about 2, can be used. Among the anion exchangers, the diethyl-aminoethyl (DEAE) group has a pK_a of about 10 and so is usable at pH values of 9 or less, whereas the strong anion exchanger quaternary methyl-lammonium is charged at all pH values.

Dextran is a polymer of glucose formed by fermentation of sucrose by the micro-organism *Leuconostoc mesenteroides;* it is marketed under the trade name **Sephadex**. **Agarose** is a polymer of the repeating unit agarobiose, which consists of D-galactose and 3,6-anhydrogalactose; it is marketed under the trade name of **Sepharose**. Cross-linked polyacrylamide gels are also sometimes used for gel permeation chromatography.

Table 3.2 Some ion exchange groups used in protein chromatography

Exchange group	Abbreviation	Structure	Type
Carboxymethyl	CM	$-CH_2CO_2^-$	Weak cation
Propylsulfonate	SP	$-CH_2CH_2CH_2SO_3^-$	Strong cation
Diethylaminoethyl	DEAE	$-CH_2CH_2NHEt_2^+$	Weak anion
Quaternary methylammonium	Q	$-CH_2NMe_3^+$	Strong anion

To explain how ion exchange chromatography is actually carried out and how it works, it is best to give an example. To make carboxymethyl-cellulose (CM-cellulose), cellulose fibres are chemically modified by intro-duction of the carboxymethyl groups, mainly on position 6 of the glucose residues. A partial structure is shown in **1**, with the introduced group in brown. The material is available commercially, as are others with dif-ferent ion exchange groups or with different supporting materials such as Sephadex.

1

The ion exchanger is packed into the chromatography column and equilibrated with a buffer of suitable strength and pH. A good choice might be sodium acetate buffer of concentration 0.01 mol dm^{-3} and pH 5.50. Under these conditions the carboxymethyl groups will be fully deprotonated and neutralized by sodium ions (Figure 3.3). The protein solution is equilibrated to the same conditions either by **dialysis** or by gel filtration (see Section 3.3.6) and run onto the column. Proteins in the mixture will be of three types. Some will be positively charged, some negatively charged and some will be isoelectric and have no net charge. Proteins in the second and third categories will have no affinity for the ion exchange groups and will pass straight through the column on washing with the starting buffer. Positively charged proteins, however, will exchange with sodium ions associated with the carboxymethyl groups (hence the name **ion exchange chromatography**) and will become bound to the matrix.

> In dialysis, the protein solution is contained inside a bag composed of a **semi-permeable** material; that is, one that will allow the passage by diffusion of small molecules but not of large ones. If the bag is placed in a large excess of a buffer solution, then the buffer components will move across the semi-permeable membrane until the concentrations of the components on the two sides are (nearly) equal.

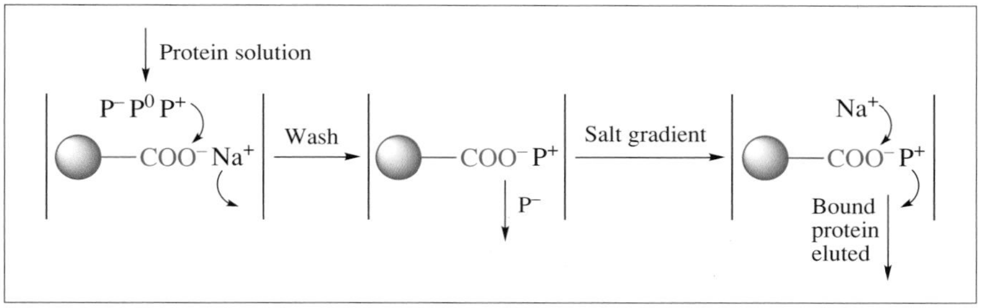

Figure 3.3 Ion exchange chromatography using CM-cellulose. A single bead of the exchanger is shown in brown. P^0, P$^+$ and P$^-$ represent isoelectric, net positively charged and net negatively charged proteins, respectively

It remains to elute the bound proteins from the column. The easiest way to do this is to pass buffer through the column containing an increasing concentration of a salt such as sodium chloride. As the sodium ion concentration increases, sodium ions will compete with protein for binding sites on the exchanger and the proteins will be displaced in order of increasing positive charge.

Figure 3.4 shows an idealized elution profile from such an experiment. The first peak was eluted with the starting buffer. After that a salt gradient was applied to elute bound proteins. In practice, if the mixture of proteins is complex, the peaks will overlap. An actual example is shown in Figure 3.5.[4] In this work it was intended to purify two enzymes, called aspartate aminotransferase and malate dehydrogenase, from pig heart muscle. A partially purified protein extract containing both enzymes was applied to a column of CM-Sephadex at pH 6.8 and the unabsorbed protein was eluted (peak not shown). A salt gradient was then applied. The dashed line shows the concentration of protein in each fraction as measured from the absorbance at 280 nm. The solid lines show the results of assay of each fraction for the activity of the two enzymes. It can be seen that essentially complete separation of the two was achieved. Neither

enzyme was pure at this stage (as can be deduced from the elution profile since neither peak of enzyme activity coincides with a discrete protein peak), but the fractions containing each enzyme could be combined
and subjected to further purification.

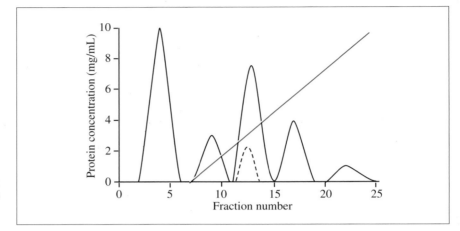

Figure 3.4 Elution profile from ion exchange chromatography. The brown solid line represents the concentration of sodium chloride in the gradient applied at fraction 6. The dashed line shows the concentration of the protein to be isolated

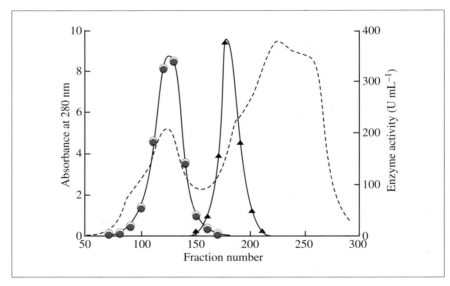

Figure 3.5 An example of an elution profile from chromatography on CM-Sephadex. Dashed line: absorption at 280 nm; circles: aspartate aminotransferase activity; triangles: malate dehydrogenase activity. (Redrawn from Barra et al.[4] with the permission of Blackwell Publishing)

Worked Problem 3.4

Q Refer to the elution profile in Figure 3.5. Which of the two
enzymes, malate dehydrogenase or aspartate aminotransferase, is
the more basic? What would you expect to happen if the two
enzymes were subjected to ion exchange chromatography on
DEAE-cellulose at pH 6.8?

A The malate dehydrogenase is more strongly attached to the CM-Sephadex since it is eluted at the higher salt concentration. This means that it is more positively charged than is aspartate aminotransferase and hence more basic. Since both of the enzymes are positively charged at pH 6.8, neither of them would be absorbed onto DEAE-cellulose at this pH.

3.3.4 Hydrophobic Interaction Chromatography

The stationary phase for this form of chromatography comprises inert beads coated with a layer of hydrocarbon chains (usually butyl, octyl or phenyl), and absorption to the support occurs between these hydrophobic chains and the hydrophobic regions that are found on the surfaces of some proteins. These regions are surrounded by a layer of structured water (see the discussion of hydrophobicity in Section 1.3.3) which tends to mask the hydrophobic residues, but this layer can be stripped off by the addition of a high concentration of a salt. Hence, typically the column is equilibrated with a buffer containing 2.0 mol dm^{-3} ammonium sulfate and the protein sample is adjusted to the same conditions. On application of the sample, those proteins with hydrophobic surfaces will bind to the support whereas hydrophilic proteins will pass through unabsorbed.

To complete the chromatography it is necessary to elute bound proteins, and the first thing to try is application of a gradient of deceasing ammonium sulfate concentration. This should be effective in eluting weakly or moderately hydrophobic proteins. A problem arises, however, if the protein of interest is strongly bound to the column. In such cases the protein will either leak off slowly in a large volume of solution, or it may not elute at all. In the later case, the only recourse is to elute with increasingly hydrophobic agents such as alcohols or non-ionic detergents, or, in the last resort, with solutions containing **chaotropic agents** such as guanidinium thiocyanate. These will definitely elute the protein, but will almost certainly lead to partial or complete loss of biological activity.

In summary, hydrophobic chromatography is a useful technique if the protein of interest is only moderately hydrophobic. It also has the virtue that it can be used straight after an initial purification step by salt fractionation without the need to remove the residual ammonium sulfate that is always present when the precipitated protein is re-dissolved. All too often, however, a protein subjected to hydrophobic chromatography is never seen again – at least not in an active form!

Chaotropic agents are molecules such as urea, guanidinium chloride and guanidinium thiocyanate that decrease hydrophobic interactions, probably by modifying the structure of water. At high concentrations they lead to denaturation of proteins.

3.3.5 Affinity Chromatography

The general principles of affinity chromatography are illustrated in Figure 3.6. The essential feature is that a ligand which is specific for the protein of interest is attached to a solid support. This is shown diagrammatically as a triangular-shaped ligand attached to the support, whereas the sample contains proteins that have binding sites for triangles, circles and squares. The protein with the binding site for triangles will attach to the ligand, whereas the others will have no affinity for the ligand and will pass straight through the column. It then remains to elute the bound protein by passing through the column a solution containing an excess of the free ligand. Of course, in practice, the recognition is not based simply on shape. It will be a combination of geometry plus a range of interactions – electrostatic, hydrophobic, hydrogen bonding, and so on – between the ligand and the binding site. Some examples will help to make this more concrete, as well as illustrating some of the problems that are encountered in affinity chromatography.

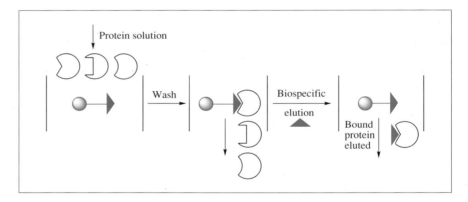

Figure 3.6 Principles of affinity chromatography

Enzyme Affinity Chromatography

The highly selective binding of an enzyme and its substrate is just what is required for affinity chromatography. In this case the substrate, or a closely related molecule termed a **substrate analogue**, is bound to the solid support.

Let us take a specific example. Fumarase catalyses reversible hydration of fumarate (*trans*-butenedioate) to yield malate (2-hydroxybutanedioate) (Scheme 3.3). An affinity matrix for this enzyme can be made by linking pyromellitic acid (benzene-1,2,4,5-tetracarboxylic acid, **2**) to a solid support *via* one of its carboxylate groups, to yield an affinity matrix as shown in **3**. If a sample containing fumarase is loaded onto a column of this material, then the fumarase binds to the immobilized substrate

This is a good example of the stereospecificity of enzymes. Fumarase acts only on fumarate and not at all on the *cis* isomer (which is called maleate). Similarly, it produces 100% of the isomer of malate shown and none of the enantiomer.

analogue whereas some of the contaminating proteins pass straight through. The enzyme can be eluted by passing a solution containing malate through the column.[5]

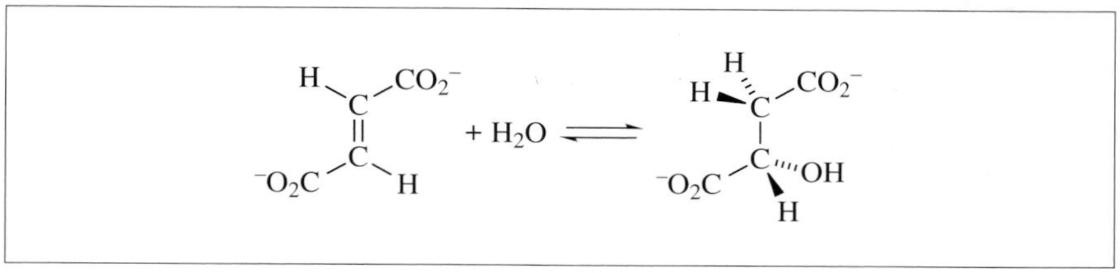

Scheme 3.3

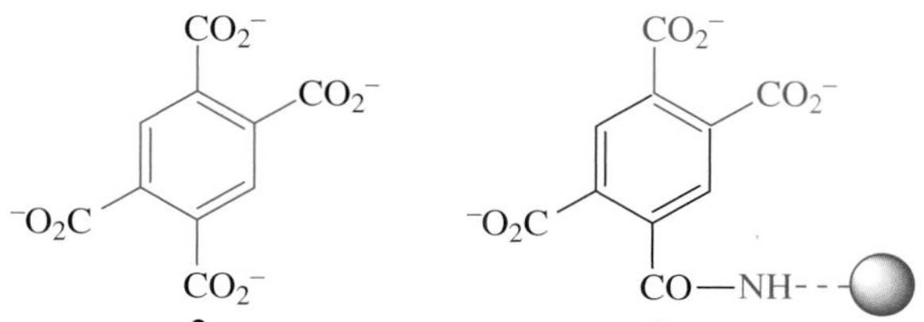

There are some interesting points about this example. Firstly, other proteins beside fumarase bind to the affinity matrix, and so a pure sample of the enzyme is not obtained. The main reason for this is that the ligand is negatively charged and so the affinity matrix will also bind positively charged proteins by purely electrostatic forces. This effect can be reduced by working at as high a pH as practical to reduce the number of positively charged proteins in the mixture. The other way to improve the specificity is to use **biospecific elution**; that is, to elute the enzyme with a low concentration of a substrate or substrate analogue rather than to use general elution with salt. Malate serves this purpose in the present example. Indeed, it is one of the tests of genuine affinity chromatography that elution can be achieved biospecifically.

The other interesting thing about this example is that it is not immediately apparent why it works! Presumably the enzyme recognizes the part of the affinity label coloured brown in **3**. However, note that this is not really a structural analogue of the substrate. To start with, the carboxylate groups in fumarate are *trans* whereas they are *cis* in **3**. Also, of course, the bonds in the aromatic ring are not really alternating double and single, even though they are usually drawn that way; benzene is aromatic. Perhaps the affinity label resembles more closely the transition state of the reaction in Scheme 3.3 than it does the ground state. The fact of the matter is that affinity chromatography does work, and emphasizes that the protein chemist has to be inventive in designing potential affinity ligands for any purification.

Dye Ligand Chromatography

Affinity chromatography using triazine dyes such as **4** (known as Cibacron Blue) was discovered by accident when it was observed that the dye had the property of binding to some proteins but not to others. The proteins that bind the dye are generally enzymes that use **nucleotide cofactors** such as adenosine triphosphate (ATP), nicotinamide adenine dinucleotide (NAD$^+$) and Coenzyme A (CoA). These molecules have in common the partial structure shown in **5**, and it is likely that the sites on these enzymes that bind this structure are the sites that recognize the triazine dyes. Triazine dyes are easily attached to solid supports such as dextran by nucleophilic displacement of the chlorine atom, and the resulting material can then be used as an affinity matrix. Elution of bound proteins can be achieved by increasing the salt concentration in the buffer or, better, biospecifically using a solution of the nucleotide for which the enzyme of interest is specific.

Cofactors are molecules that participate with enzymes in carrying out particular classes of reaction. An example is NAD$^+$, which is the cofactor for oxidative enzymes called **dehydrogenases**. Typical is alcohol dehydrogenase, which oxidizes ethanol to acetaldehyde (ethanal, Scheme 3.4). This is the first step in the process by which we remove ethanol from the body.

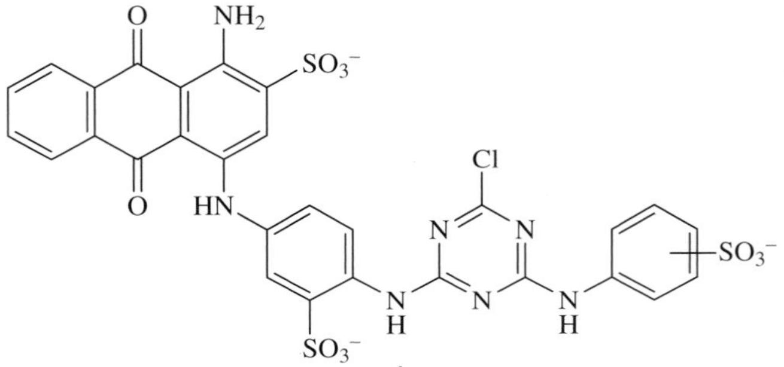

4

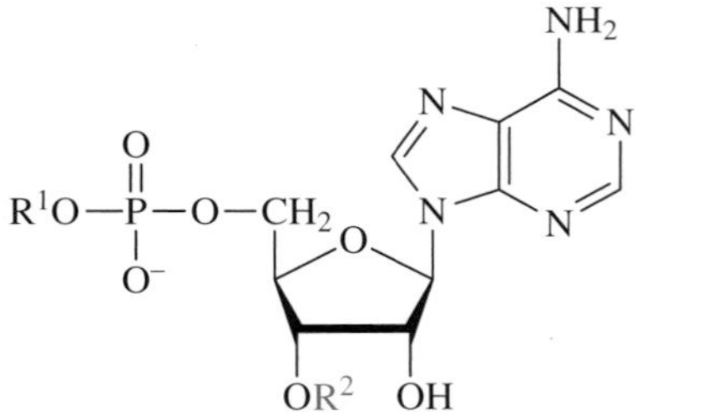

ATP and NAD$^+$, R^2 = H; CoA, R^2 = PO$_3^{2-}$

5

$$\text{MeCH}_2\text{OH} + \text{NAD}^+ \longrightarrow \text{MeCHO} + \text{NADH} + \text{H}^+$$

Scheme 3.4

A range of triazine dyes has been developed and these have been found to have different selectivities for the various nucleotide-dependent enzymes. It is sometimes useful to use more than one of these matrices sequentially for purification of a particular enzyme. A particularly successful example of this approach has been described by Worrall[6] for the purification of the enzyme CoA synthase. A crude extract from pig liver was first subjected to affinity chromatography using Procion Red Sepharose (Procion Red is a triazine dye with a somewhat more complex structure than that of Cibacron Blue). The enzyme was retained on the column and was subsequently eluted by applying a linear gradient of KCl. The elution profile is shown in Figure 3.7A. The fractions containing enzyme activity were pooled and applied to a column of Blue Sepharose. Again the enzyme was retained by the column and, after washing the column to remove impurities, was eluted by applying a solution containing 0.1 mmol dm^{-3} Coenzyme A. The product of the second step was analysed by **SDS-PAGE** (Figure 3.7B) and shown to be essentially pure.

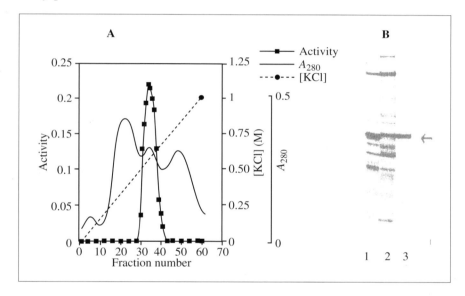

Figure 3.7 Purification of CoA synthase. **A** Elution profile; **B** SDS-PAGE (Redrawn from Worrall[6] with the permission of Humana Press)

Box 3.5 Sodium Dodecyl Sulfate Polyacrylamide Gel Electrophoresis

Electrophoresis means the movement of charged species in an electrical field, and electrophoresis in gels is the most powerful technique available for the analysis of protein mixtures. The most commonly used gel for work with proteins is **polyacrylamide**. In the original form of the technique, the protein mixture to be analysed is applied to a polyacrylamide gel at a suitable pH and an electrical field applied. After a suitable time the gel is stained with a dye which binds to proteins but not to the gel; the usual stain is called Coomassie Blue. The result is a pattern of bands each representing at least one, but possible more, of the proteins in the mixture.

Interpretation of the results of this form of electrophoresis is complicated because the rate at which a protein moves depends on two factors. Firstly, the more highly charged a protein is, the more rapidly it will migrate. Secondly, the gel has a sieving effect – the proteins have to get through the gel particles – so the smaller the protein, the faster it will migrate. This means that a large, highly charged protein and a small, less charged protein might migrate at the same rate. This problem is overcome in sodium dodecyl sulfate polyacrylamide gel electrophoresis (invariably known as **SDS-PAGE**).[7] Sodium dodecyl sulfate is an anionic detergent. It binds to proteins, and in so doing destroys their native three-dimensional structures, converting them into rigid rod-like molecules. A large amount of detergent is bound (about 1.4 g/g of protein) and this makes the protein strongly negatively charged irrespective of its amino acid composition. Hence if proteins are subjected to electrophoresis in the presence of SDS, they all migrate to the anode at rates which depend only on their sizes (the smaller the protein, the faster it migrates). Except for the circumstance where two proteins are the same size, or very nearly the same size, they will separate on SDS-PAGE. Or, putting it the other way round, the appearance of a single band on the gel after SDS-PAGE provides very strong evidence that the protein sample is pure.

The electrophoretic patterns of three samples are shown in Figure 3.7, with an arrow marking the position of the pure enzyme. Lane 1 is the crude sample that was applied to the first column. Note that the desired enzyme is present at such a low level in the crude extract that it is not visible on the gel; the large band running just faster than the enzyme is an impurity. Lane 2 shows the active fraction from the column of Procion Red. The enzyme of interest is now clearly visible but there are several contaminants present, some of which were not visible in the crude sample. Lane 3 shows the pure enzyme as eluted from the Blue Sepharose.

Immobilized Metal Affinity Chromatography

Some proteins have binding sites for metal ions, typically Ca^{2+}, Zn^{2+}, Cu^{2+}, Ni^{2+} or Fe^{2+}. These metal-binding sites can be exploited in **immobilized metal affinity chromatography** (known as **IMAC**). In this technique a metal chelator such as iminodiacetate is coupled to the solid support and loaded with metal ions. The resulting structure is shown in **6**, where the X groups represent coordinated water and M is the chosen metal ion.

The mixture of proteins to be purified is first treated with a chelator to remove bound metals from the proteins, and then dialysed to remove excess chelator. When the protein mixture is loaded onto the column, those proteins that have a binding site for the metal immobilized on the column will displace the coordinated water molecules and become bound to the metal. Other proteins will pass through as the column is washed with the starting buffer. Absorbed proteins can then be eluted by passing a buffer containing an excess of the metal ion through the column.

6

Worked Problem 3.5

Q The enzyme carboxypeptidase has a zinc ion at its active site. Suggest how this fact might provide a useful step in purification of the enzyme. Comment on any possible limitations of the method chosen

A This is a case where IMAC should be useful. After initial extraction and, probably, partial purification by ion exchange chromatography, the zinc would be removed by treatment with a chelating agent and the protein mixture passed through a column of immobilized iminodiacetate loaded with zinc. After washing off unabsorbed protein, the carboxypeptidase would be eluted with a buffer containing zinc ions. The likely limitation is that zinc is a relatively commonly occurring metal ion in proteins, and so the product would probably contain a mixture of such proteins.

Lectin Affinity Chromatography

This, and the final example of affinity chromatography described below, is a little different from the cases described before. In those previous cases the ligand which is recognized by the protein is bound to the solid support. In the remaining two cases it is the species joined to the solid support that recognizes the protein.

As outlined in Section 3.2, many proteins, particularly ones which are components of biological membranes, are glycoproteins. The carbohydrate chains are added to the protein after completion of synthesis of the polypeptide chain, and the precise structure of the polysaccharide is specific for the particular protein. Lectin affinity chromatography provides a method of separating proteins of a particular carbohydrate class from other classes and from non-glycosylated proteins.

Lectins from different sources differ in their specificities. A few examples are given in Table 3.3. The most commonly used is concanavalin A (Con A) which has specificity for α-D-mannopyranoside residues. Hence if a column is made with this lectin immobilized on a solid support, glycoproteins containing α-D-mannopyranoside residues will be specifically absorbed out from a mixture of proteins. Non-glycosylated proteins, and glycoproteins not containing α-D-mannopyranoside residues, will pass through the column unabsorbed. The retained glycoproteins can then be eluted by passing a solution containing a high concentration of D-mannose or, better, methyl-α-D-mannoside through the column. The product of such an experiment will be a mixture of glycoproteins of the particular class containing α-D-mannopyranoside residues, the components of which can then be separated by other methods.

Table 3.3 Specificities of some lectins

Name	Specificity	Eluent
Concanavalin A	α-D-Mannopyranosyl	Methyl α-D-mannoside
Wheat germ agglutinin	N-Acetyl-D-glucosamine	N-Acetyl-D-glucosamine
Jacalin	D-Galactopyranosyl	Melobiose

Worked Problem 3.6

Q You wish to isolate a glycoprotein, the polysaccharide chain of which is known to contain N-acetyl-D-glucosamine residues. Suggest a step that would be included in the purification.

A Chromatography using immobilized wheat germ agglutinin.

Immunoaffinity Chromatography

In this method an antibody raised against a particular protein is coupled to a solid support. The affinity matrix so produced can then be used to absorb that protein specifically from a mixture of other proteins. Because of the extreme selectivity of antibody–antigen reactions, the only protein that should be bound to the affinity matrix is the protein to which the antibody is raised.

Box 3.6 Coupling of Antibodies to Solid Supports

There are several ways in which this can be done. One of the most common involves activating the polysaccharide support with cyanogen bromide to produce a reactive intermediate, probably a cyclic imide, which will react with ligand amino groups to form an isourea derivative (Scheme 3.5). In the case of antibodies, the ligand amino groups are provided by the side chains of the amino acid lysine. Cyanogen bromide is a extremely nasty substance and has to be handled with great care!

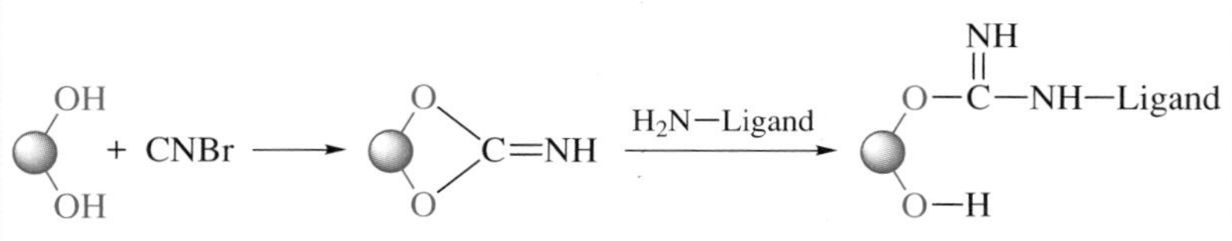

Scheme 3.5

An alternative is to use commercially available materials called Affi-Gels. These are *N*-hydroxysuccinimide esters that react with amino groups to form amide linkages (Scheme 3.6). The reactive ester is attached to a spacer arm, which in the example shown is 10 atoms long. The advantage of this is that the reactive group is some distance away from the solid support, and so the possibility of a low coupling efficiency due to steric interaction between the gel and the antibody is minimized.

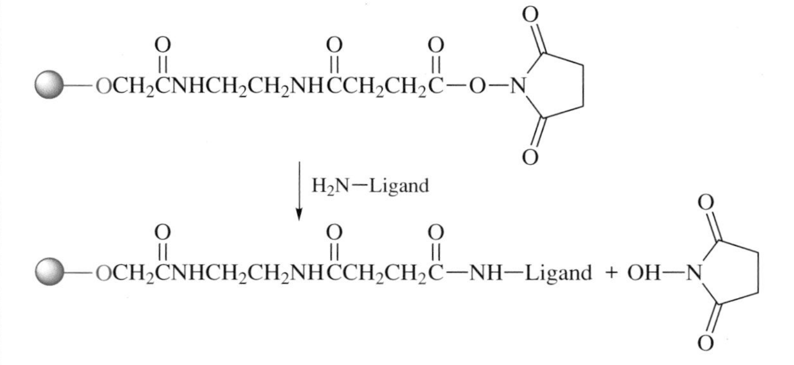

Scheme 3.6

It seems that immunoaffinity chromatography should provide the ideal method for purification of a protein. In principle, once the affinity matrix has been made, it should be possible to isolate the protein from a crude mixture in a single experiment.

There are two problems. Firstly, there is the problem of obtaining the antibody in the first place. As mentioned previously, antibodies are produced when a protein from one organism is injected into the blood stream of a different animal. Obviously it is no use injecting a mixture of proteins into the animal because the antiserum produced will contain antibodies to all the proteins in the mixture. This means that the protein of interest will have to be purified first by some other set of procedures so that it can be used to raise the antibody. This is not quite as strange as it might seem because very often it is necessary to purify a protein on many occasions, for example if it is to be used for an extensive series of investigations, or if it is to be sold as a therapeutic agent. In such cases it is worthwhile developing a method to purify the protein by other methods, raise an antiserum, and then make an affinity column which can be used repeatedly for subsequent purifications.

Box 3.7 Monoclonal Antibodies

There is an alternative approach which does not require use of a pure antigen. A mixture of proteins is injected into a mouse and, several weeks later, its spleen is removed and the lymphocytes (antibody-producing cells) recovered from it. These lymphocytes are then fused (combined with) a special class of cells called **myeloma cells**; these cells have the property of dividing uncontrollably. Each of the resulting **hybrid** cells, called **hybridomas**, can be grown in culture to produce a **clone**, and each clone will secrete a single type of antibody molecule. In remains only to identify the clone which produces antibodies to the antigen of interest and we then have a factory for producing an essentially unlimited amount of the specific antibody required. These are referred to as **monoclonal antibodies**. Monoclonal antibody technology is of enormous importance in science and medicine and its inventor, César Milstein, was awarded the Nobel Prize in Medicine for this work in 1984.

The other problem is more severe. Antibodies bind antigens very strongly; it is important for their biological role that they should do so. As a result of this, it is usually very easy to get the protein of interest bound onto the affinity matrix, but is often extremely difficult to get it off again. The protein is typically loaded onto the matrix at a pH of about 8.5 and, after unbound protein has been washed off, the pH is lowered to attempt to elute the bound protein. There is, however, a limit to how far one can go. All proteins are unstable and lose biological activity at a low enough pH; for most proteins the safe limit might be pH 2–3. This may well not work because even at these low pH values the binding of the antigen to the antibody might still be very tight. If that fails, then elution can be attempted using solutions of urea or even of a chaotropic agent. These will almost certainly elute the protein, but quite likely it will be in an inactive from.

In summary then, immunoaffinity chromatography is a very valuable technique if:

(1) it is intended to repeat the purification sufficiently often to make it worthwhile going to the trouble of making an antibody;
(2) an antibody is obtained that binds the antigen sufficiently strongly to retain it on the column under the starting conditions, but not so strongly that the protein cannot be released under conditions where the biological activity is retained.

3.3.6 Gel Permeation Chromatography

In gel permeation chromatography (also known as gel filtration, or size exclusion chromatography) the protein mixture is passed through a column of porous beads. The beads are made of cross-linked polysaccharide chains and the degree of cross-linking can be varied so as to vary the sizes of the pores. When the beads are suspended in buffer solution and packed into the column, there will be solvent both inside the beads and also filling the interstices between them. What happens when a mixture of proteins is passed through the column is shown diagrammatically in Figure 3.8.

Figure 3.8 Principles of gel permeation chromatography. Open circles are the gel beads and brown objects are protein molecules (not on the same scale). Small proteins have access to pores in the beads and will pass through slowly; large proteins do not and will emerge first

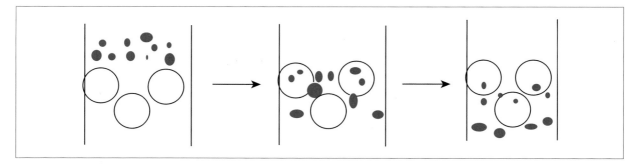

Consider first a protein which is large compared with the size of the pores and hence cannot enter them. It has access only to the buffer between the beads (about 25% of the total liquid volume of the column) and hence will emerge from the column very quickly. A very small protein, on the other hand, can freely enter the pores and has access to the total liquid volume of the column. It will, therefore, emerge later. An intermediate sized protein will have partial access to the pores and will emerge in an intermediate volume of buffer.

This is all very well for a simple mixture of three proteins with very different sizes, but in fact the resolving power of the method is rather poor. The peaks of retarded proteins tend to be rather broad – after all, the method depends on the diffusion of proteins into and out of the beads – and hence separation of proteins with only small size differences is not possible. Although gel permeation media are available covering a wide range of size exclusion limits, the method is of limited application in protein purification.

That being said, gel permeation chromatography is very useful as an alternative to dialysis for removing salts from protein solutions, or for changing the buffer. For example, if a sample of protein solution in buffer A is passed through a gel permeation column equilibrated in buffer B and from which the protein is excluded, the protein will move ahead of the buffer in which it was applied and will emerge in buffer B. Changing buffers is a frequent requirement in protein purification.

3.4   A Typical Purification

We have now looked at all the major techniques available for protein purification and the question arises as to how to put them together in practice. The answer is that there is no general rule: each purification problem is unique. However, a general approach for the isolation of a protein at relatively large scale (perhaps starting with a few kilograms of tissue) would be:

- Homogenization of the tissue in buffer and removal of insoluble material by centrifugation
- Initial crude fractionation and volume reduction by fractional precipitation with salt
- Remove residual salt and change the buffer by dialysis
- Large-scale (column volume of, say, 2 dm³) ion exchange chromatography
- Change buffer by dialysis or gel permeation chromatography
- Second ion exchange chromatography step under different conditions (*e.g.* using an exchanger of opposite charge)
- Change buffer by dialysis or gel permeation chromatography
- Affinity chromatography to exploit special properties of the protein

The steps are ordered to move from the more general to the more specific, and from ones that are easy to do on a large scale to ones that are only practical on a small scale. At each step it will be necessary to carry out trials to find the optimum conditions for the step, and to analyse the product by electrophoresis to see how the purification is progressing. One factor which is common to all purifications is that it is essential to have a method available for measuring the amount of the desired protein in the fractions at each stage of the purification so that the yield can be determined. A step resulting in a low yield will have to be abandoned and replaced by an alternative, otherwise the final yield of the process may be unacceptably low. Five purification steps each with a step yield of 50% will lead to a final overall yield of only 3%!

A typical purification of an enzyme starting from 10 kg of pig liver is summarized in Table 3.4, taken from work done in the author's own laboratory. Quantitation was easy here because the catalytic activity of the enzyme could be determined. Enzymologists usually define the activity of an enzyme in terms of **Units** where 1 U is the amount of enzyme that produces 1 µmol of product per min under standard assay conditions. The **specific activity** is then the number of U/mg of protein. The requirements for a useful step in the purification are that the yield should be high, and that there should be a substantial increase in the specific activity. The **purification factor** is given by the ratio of the specific activity of the enzyme at each step compared with that of the original homogenate. In the present case the purification factor achieved was about 6000, which means that the enzyme was present at a level of only about 0.017% of the protein in the initial homogenate. Note how the scale of the procedure decreases from beginning to end. It started with 9 dm^3 of homogenate and ended with 12 cm^3 of pure protein solution. Similarly, 28 mg of pure enzyme were obtained from 340 g of protein in the extract.

Table 3.4 Example of a purification table

Fraction	Volume (cm^3)	Protein concentration (mg/cm^3)	Total protein (mg)	Activity (U/cm^3)	Total activity (U)	Specific activity (U/mg)	Purification factor	Overall yield (%)
Homogenate	9000	37.8	340,200	1.8	16,200	0.048	1	100
45–65% (NH$_4$)SO$_4$	530	190.0	100,700	23.5	12,455	0.124	2.6	77
CM-cellulose	420	19.5	8,190	25.5	10,710	1.31	27.2	66
DEAE-Sephadex	75	7.5	563	138	10,350	18.4	383	64
Affinity chromatography	12	2.33	28	683	8,200	293	6100	51

Box 3.8 The Contribution of Genetic Engineering to Protein Purification

Modern methods of molecular biology allow the gene coding for a particular protein, perhaps a human protein, to be inserted into a bacterial cell; this technology is generally known as **genetic engineering**. The bacteria, grown in culture, can then be used as a source of the foreign protein. This is particularly useful if the original source of the protein is hard to obtain (as human tissue certainly is) or if the protein is present in the original source at unusually low levels. Both of these conditions apply, for example, to proteins called **interferons**. These proteins are used in cancer therapy and the only practical way of obtaining them is from genetically engineered micro-organisms.

It is possible to go further. Before the gene for the protein of interest is inserted in the bacterial cell, it can be modified in such a way as to introduce a new characteristic in its protein product which will make the protein easier to purify. For example, the gene can be modified in such a way that the protein product terminates with a string of histidine residues. The point of this is that the so-called poly-His tail has strong metal-ion binding properties and so the protein can be purified using IMAC. As another example, it is possible to add to the C-terminus a small peptide for which an antibody is available. The antibody can then be used to purify the protein by immunoaffinity chromatography. Even more cunning, if it is arranged that the bond between the natural protein and the added small peptide can be broken easily, perhaps by an enzyme, then the required protein can be detached from the antibody under mild conditions, thus avoiding the major difficulty often encountered with immunoaffinity chromatography. All this may sound a lot of work, but it is very worthwhile when a valuable chemotherapeutic product is the result!

Summary of Key Points

- Protein purification is difficult because any individual protein usually constitutes less (often much less) than 1% of the proteins in a tissue extract. Moreover, all of the impurities are also proteins and have the same basic properties.
- To purify proteins it is necessary to exploit the properties by which proteins differ from one another.

- Most methods of protein purification involve some form of column chromatography.
- The most generally useful method is ion exchange chromatography. Other chromatographic methods exploit special properties of proteins such as hydrophobicity, specific binding sites, or the possession of surface carbohydrate chains.
- Immunoaffinity chromatography is, in principle, the most specific method available for protein purification. It sometimes suffers from the disadvantage, however, that proteins once bound to the antibody may be difficult to detach.
- It is important at all stages during a purification to keep track of the yield of the protein of interest. Low step yields are not acceptable because in a multi-step process they will lead to low overall yields.
- The most useful analytical technique to test the purity of a protein is SDS-PAGE.

Problems

3.1. What would be the pH of a buffer obtained by adding 0.025 mol of NaOH to 1 dm^3 of 0.100 mol dm^{-3} NaH$_2$PO$_4$ (sodium dihydrogen phosphate), given that the second pK_a of phosphoric acid is 7.20?

3.2. How much HCl would you need to add to 1 dm^3 of a 0.100 mol dm^{-3} solution of a weak base with a pK_a of 8.00 to obtain a buffer of pH 7.50?

3.3. A paper reports that, during salt fractionation, the protein was collected by centrifugation at 10,000 g for 15 min. You have a centrifuge with a rotor of radius 25 cm. What angular velocity would you need to use to replicate the reported conditions?

3.4. Three proteins have isoelectric points (*i.e.* the pH at which the net charge is zero) of 6.0, 7.0 and 8.0. You wish to separate them. How might you proceed to do so using chromatography on DEAE-cellulose? From the information given, is it possible to predict how the pure proteins would behave on SDS-PAGE?

3.5. Refer to Box 3.8. Another way in which proteins have been genetically engineered to make their purification easier is by the addition of strings of 8–10 arginine or lysine residues at the C-terminus. What technique would you use to capitalize on this modification (remember that lysine and arginine are basic amino acids)?

3.6. When a protein is precipitated with ammonium sulfate, collected by centrifugation, and re-dissolved in buffer, the resulting solution contains a substantial amount of residual ammonium sulfate. (a) Name a purification technique for which the presence of this ammonium sulfate is an advantage. (b) Name two methods that you could use to get rid of it.

3.7. Complete the purification table in Table 3.5. Why would this be considered a poor purification schedule? Where did it go wrong?

Table 3.5

Fraction	Volume (cm³)	Protein concentration (mg/cm³)	Total protein (mg)	Activity (U/cm³)	Total activity (U)	Specific activity (U/mg)	Purification factor	Overall yield (%)
Homogenate	1000	42.5		0.9				
45–65% $(NH_4)SO_4$	120	134.0		6.8				
Hydrophobic chromatography	230	3.7		0.8				
DEAE-Sephadex	55	0.92		2.4				
CM-Sephadex	15	0.48		7.2				

References

1. S. Doonan, in *Protein Purification Protocols*, ed. S. Doonan, Humana Press, Totowa, NJ, 1996, p. 38.
2. I. Roitt, J. Brostoff and D. Male, *Immunology*, 6th edn., Mosby, St. Louis, 2001.
3. S. Doonan, in *Protein Purification Protocols*, ed. S. Doonan, Humana Press, Totowa, NJ, 1996, p. 1.
4. D. Barra, F. Bossa, S. Doonan, H. M. A. Fahmy, F. Martini and G. J. Hughes, *Eur. J. Biochem.*, 1976, **64**, 519.
5. M. C. O'Hare and S. Doonan, *Int. J. Biochem.*, 1985, **17**, 279.
6. D. M. Worrall, in *Protein Purification Protocols*, ed. S. Doonan, Humana Press, Totowa, NJ, 1996, p. 169.
7. U. K. Laemmli, *Nature*, 1970, **227**, 680.

Further Reading

E. L. V. Harris and S. Angal, *Protein Purification Methods*, IRL Press, Oxford, 1989.

E. L. V. Harris and S. Angal, *Protein Purification Applications*, IRL Press, Oxford, 1990.

S. Doonan, in *Encyclopedia of Separation Science*, ed. I. D. Wilson, E. R. Adlard, M. Cooke and C. F. Poole, Academic Press, London, 2000, p. 4547.

4

Determination of the Covalent Structures of Peptides and Proteins

Aims

By the end of this chapter you should understand:

- How the relative molecular mass and amino acid composition of a protein are determined
- What is meant by the term quaternary structure and how to discover whether a particular protein has this level of structure
- How the nature of the amino acid at the N-terminus of a peptide or protein can be determined
- How the amino acid sequence of a peptide can be determined by sequential degradation from the N-terminal end
- How amino acid sequencing by chemical methods is extended to proteins
- How the positions of disulfide bridges are established
- The contributions that mass spectrometry is making to determination of the structures of peptides and proteins

4.1   Introduction

We know that peptides and proteins are made from amino acids linked together by peptide bonds, and so determination of the covalent structure amounts to discovering the order in which the residues occur in the chain; that is, the amino acid sequence (or **primary structure** as it is commonly called). Most of what follows refers particularly to proteins which, because of their greater size and extra level of structural complexity, present more of an analytical challenge. The exception is in the description of sequence analysis, where peptides are dealt with first, followed by what is required to extend the methods to proteins.

Before embarking on determination of the primary structure there are certain other pieces of information that are required. Firstly, we need to know the **relative molecular mass** (M_r) of the protein, which defines the magnitude of the sequencing problem. It is also necessary to know whether or not the protein consists of a single polypeptide chain. There are two different circumstances where a protein can consist of two, or more, chains. Firstly, but relatively uncommon, the protein may have different polypeptide chains linked by disulfide bridges. Insulin, as we have seen, is a case in point. It consists of two chains, one of 21 amino acids and the other of 30, linked by two disulfide bridges. Much more common, the native protein may consist of two or more chains that are associated together, but not covalently linked. This is referred to as **quaternary structure** (we will see what secondary and tertiary structures are in Chapter 5). Usually the individual chains, referred to as **monomers**, have the same amino acid sequence, so this is not a problem. If, as sometimes occurs, the protein consists of two monomers of different primary structure, then the chains have to be separated before analysis of their sequences.

The next thing to determine is the amino acid composition of the protein, or of its constituent polypeptide chains. This information is not strictly necessary for amino acid sequence analysis, but is useful for a variety of reasons that will be discussed later. With this preliminary information to hand, the sequence can be determined. Two approaches to sequence analysis will be described. One is the now classical chemical method in which the polypeptide chain is degraded one residue at a time from the N-terminus, and the residue released at each stage is determined. The other is a more recent approach that depends on the use of mass spectrometry.

One aspect of protein structure analysis that will not be covered is determination of the structures of the polysaccharide units of glycoproteins. This is a difficult and specialized area of chemistry, and is outside the scope of this book. The interested reader can find a detailed account of this topic in Chapters 97–111 of Walker.[1] This reference also contains detailed treatments of many of the topics covered in this chapter.

The average **residue relative molecular mass** (the M_r of the amino acid minus 18 for the peptide bond) in proteins is between 110 and 120. So the approximate number of residues can be obtained by dividing the protein M_r by these numbers. If $M_r = 50,000$, then the number of residues is between 415 and 455.

4.2 Relative Molecular Mass and Subunit Composition

Commonly used methods for the determination of the relative molecular masses (M_r) of proteins are gel permeation chromatography or SDS-PAGE, both of which have been covered in other contexts in Chapter 3. These are **empirical** methods. That is, they work, but they have no theoretical underpinning. In the case of gel permeation chromatography, a column is **calibrated** by passing through it a series of proteins of known

M_r and measuring the elution volume of each; that is, the volume of solution passed through the column from the point of application of the sample until the peak maximum of the eluted protein appears. It turns out that if the elution volume is plotted against the logarithm of M_r, then a smooth curve is obtained, as shown in Figure 4.1. If a protein of unknown M_r is now passed through the column and the elution volume measured, its M_r can be obtained from the calibration curve.

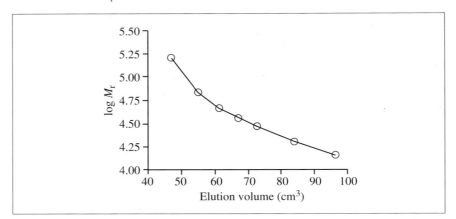

Figure 4.1 Plot of M_r against elution volume for gel permeation chromatography

Worked Problem 4.1

Q Gel permeation chromatography was carried out on a set of proteins of known M_r. The elution volumes and the M_r values are given in Table 4.1. A newly isolated protein had an elution volume from the same column of 58.0 mL. Estimate its M_r.

Table 4.1 Relative molecular masses of some proteins and their elution volumes from gel permeation chromatography

Protein	Relative molecular mass	Elution volume (cm³)
Aldolase	160,000	47.0
Serum albumin	68,000	55.2
Ovalbumin	45,000	61.5
Glyceraldehyde 3-phosphate dehydrogenase	36,000	67.3
Carbonic anhydrase	29,000	72.8
Trypsin inhibitor	20,000	84.0
α-Lactalbumin	14,000	96.5

A A plot of log M_r against elution volume is shown in Figure 4.1. From this plot, an elution volume of 58.0 mL corresponds to log $M_r = 4.72$. Hence $M_r = 52,500$.

If the elution volume of the unknown protein is near one of the ends, or outside, of the calibration curve, then the experiment must be repeated using a gel with a different degree of cross-linking and a different set of standards.

Determination of M_r by SDS-PAGE is similar in principle. In this case, a gel is run with a set of proteins of known M_r and, after staining, the distance migrated by each protein is measured and plotted against log M_r. In this case a straight line relationship is found. Again, if the distance migrated by an unknown protein is measured under identical conditions, then its M_r can be obtained from the calibration curve.

Both of these methods usually provide quite accurate values of the M_r of a protein, although with glycoproteins the gel permeation method sometimes yields an underestimate because the protein is retarded in its passage through the column by interaction of the polysaccharide chain of the protein with the matrix. SDS-PAGE is the preferred method because it is very quick to do (the whole experiment can be carried out in a couple of hours) and, more importantly, because it requires very little protein; a few micrograms are sufficient.

It is, however, often instructive to carry out both types of determination and compare the results. Suppose, for example, that the value obtained by SDS-PAGE was exactly one half of that from gel permeation chromatography. The likely explanation of this finding would be that the protein is a dimer of two identical subunits. The reason for this behaviour on SDS-PAGE is that the detergent disrupts the non-covalent interactions that are responsible for formation of quaternary structure. It would be necessary to do a control experiment to ensure that the two chains are not linked by disulfide bridges. SDS-PAGE is usually carried out in the presence of the reagent mercaptoethanol (ethanethiol). This reagent reduces disulfide bridges in proteins, as shown in Scheme 4.1, and is added to ensure that the protein is able to fully unfold and adopt an extended rod-like conformation. A gel could be run in the absence of mercaptoethanol. If the protein still migrated at or near the monomeric position, then the possibility of inter-chain disulfide bridges would be ruled out. An example of how a combination of gel permeation chromatography and SDS-PAGE can give information about quaternary structure is in the following worked problem.

Scheme 4.1

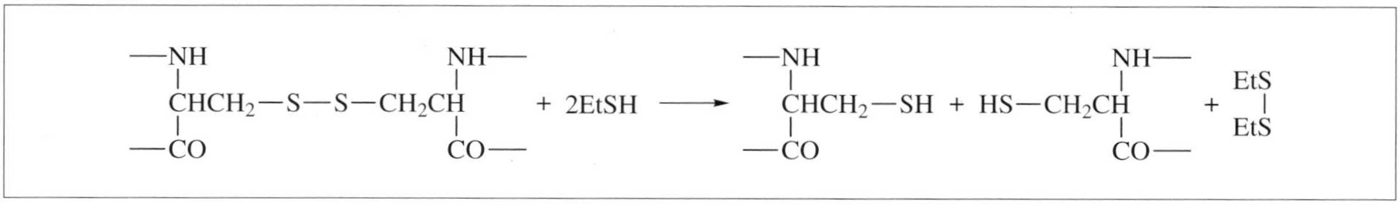

Worked Problem 4.2

Q A mixture of the same set of proteins as shown in Table 4.1, excluding aldolase, was run on SDS-PAGE in the absence of mercaptoethanol. The same unknown protein as in the previous problem was run in an adjacent track. After staining, the pattern was as shown diagrammatically in Figure 4.2. Determine the M_r of the unknown, and comment on the result.

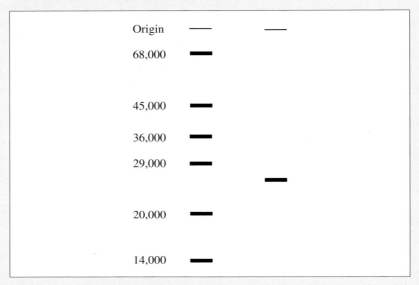

Figure 4.2 SDS-PAGE of a set of standard proteins (left) and an unknown protein (right)

Again, the degree of cross-linking of the polyacrylamide used, and the range of M_r values of the standards, will depend on the M_r of the unknown protein.

A Measure the distance moved by each standard protein from the origin and plot this distance against log M_r (plot not shown). Measure the distance moved by the unknown and, from the standard graph, evaluate its M_r. The value obtained is 26,000. This is half the value obtained by gel permeation chromatography. This suggests that the unknown protein is a dimer of two identical subunits. Since the gel was run in the absence of mercaptoethanol, we can rule out the possibility of two chains linked by disulfide bridges.

Box 4.1 Molecular Masses from Sedimentation and Diffusion Analysis

Determination of relative molecular masses by empirical methods requires the use of standards whose M_r values are already known.

This begs the question of how protein molar masses were determined in the first place. The answer is by **sedimentation** and **diffusion** analysis.

The **sedimentation coefficient** S is defined as the rate of movement of a particle through solution in a unit gravitational field. It has units of time (s). Typical values for proteins are in the region of 10^{-13} s, and this quantity is frequently referred to as a **Svedberg unit** (S).

The **diffusion coefficient** D is defined as the rate of movement in a unit concentration gradient. It has units of $m^2\ s^{-1}$.

These quantities are related to the molar mass M by:

$$M = \frac{S}{D} \times \frac{RT}{1-\bar{v}\rho} \qquad (4.1)$$

Here T is the absolute temperature, and R is the gas constant. Consideration of the units in the equation will show that if R is in $J\ mol^{-1}\ K^{-1}$, then the units of M are $kg\ mol^{-1}$. That is, it is the molar mass.

Measurement of S requires the use of an analytical ultracentrifuge, a machine developed by The Svedberg in the early 1900s.[2] The essential feature of the instrument is that observations can be made of the movement of the protein in the centrifuge cell while centrifugation is in progress. Measurement of D requires a boundary to be set up between a protein solution and the buffer in which the protein is dissolved, and observing the rate at which the protein moves across the boundary.

The quantity ρ in equation (4.1) is the density of the buffer used, and $\bar{v}$ is the partial specific volume of the protein (that is, the rate of change of volume of the solution per gram of added protein). The value of $\bar{v}$ is often obtained from the amino acid composition by summing contributions from each residue type. This is not entirely reliable, and it is better to obtain $\bar{v}$ from accurate measurements of the densities of solutions of the protein as a function of concentration. Errors in $\bar{v}$ are serious because the value of $1 - \bar{v}$ is about 0.25, so an error of 5% in $\bar{v}$ will give a 15% error in $1 - \bar{v}$ and hence in the value of the molar mass obtained.[3]

4.3 Amino Acid Analysis

First the protein has to be degraded completely into its constituent amino acids. Peptide bonds vary considerably in their resistance to hydrolysis, depending on the particular residues involved, with bonds to hydrophobic amino acids being most resistant. Complete hydrolysis requires vigorous conditions, and 6 mol dm^{-3} HCl at 105 °C for 24 h is commonly used. Under these conditions, asparagine and glutamine are hydrolysed to the corresponding acids, and so the values subsequently obtained for aspartic and glutamic acids are composites of the values for the free acids and the amides. Tryptophan is completely destroyed, but can be analysed separately after milder hydrolysis with 3 mol dm^{-3} toluene-4-sulfonic acid. Substantial loss of cysteine and cystine will also occur because of oxidation, even if the hydrolysis is carried out in evacuated tubes. In a separate analysis, these amino acids can be quantitatively converted to the stable derivative cysteic acid (**1**) by oxidation of the protein with performic acid [peroxomethanoic acid, HC(O)–O–OH].

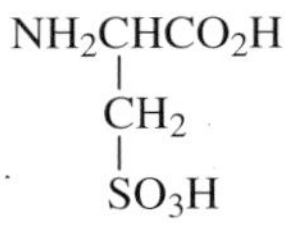

NH$_2$CHCO$_2$H
|
CH$_2$
|
SO$_3$H

1

Having obtained the hydrolysate, there are two ways of proceeding to separate and quantitate the amino acids. The classical method for separation of the amino acids is by ion exchange chromatography using sulfonated polystyrene resins. The amino acids are bound to the column at a pH of about 2.2, and then subsequently eluted by gradually increasing the pH. Eluted amino acids are mixed with **ninhydrin**, which reacts to form a purple product (Scheme 4.2). The product formed is the same irrespective of which amino acid is involved, and can be measured quantitatively by its absorption at 570 nm. Proline forms an orange product which absorbs around 440 nm and is measured separately.

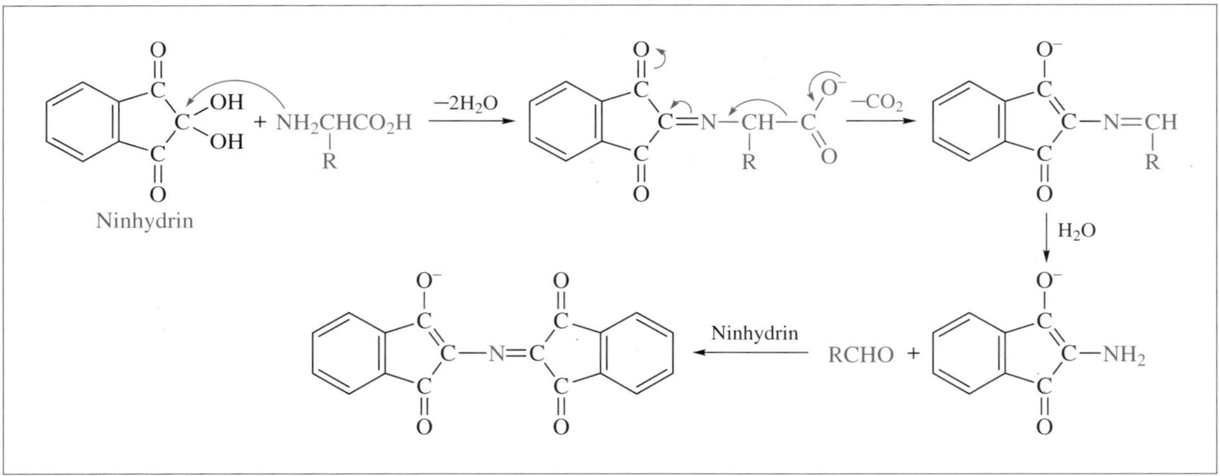

Scheme 4.2

This form of analysis was introduced by Stamford Moore and William Stein in the 1950s, and is universally referred to as the Moore and Stein method. It was subsequently automated,[4] and commercial machines are available to carry it out. After the chromatographic separation, the ninhydrin reagent is mixed with the effluent stream from the column, the solution is passed through a heater to develop the coloured product, and finally to a UV detector. The output of the detector is fed to a chart recorder. Each amino acid is identified by its time of elution and its quantity determined from the area (or height) of the peak on the chart. Automated systems have undergone extensive development over the years, particularly to increase sensitivity, and systems can be obtained which will give reliable analyses at the nmol level; that is, analysis can be carried out on 1–2 µg of protein.

The alternative is to convert the amino acids to some appropriate derivative before analysis and then to separate the derivatives using RP-HPLC (this is referred to as **pre-column derivatization**, as opposed to the Moore and Stein procedure which is a **post-column** method). A variety of reagents has been used, but they share the property that the derivatives formed are **fluorescent**. The advantage of this is that fluorescence detection can be more sensitive that UV detection, and so by using fluorescent derivatives the sensitivity of analysis can be increased about 100-fold; that is, down to the 10 pmol level. In one commercial system the amino acids are converted to the Fmoc derivatives for analysis (see Chapter 2). A more recent system[5] uses 6-(aminoquinolyl)-N-hydroxy-succinimidyl carbamate to produce the fluorescent derivatives (Scheme 4.3). Separation is carried out using a C_{18} reverse-phase column, and the separated derivatives detected and quantitated by fluorescence at 395 nm.

> Moore and Stein were awarded the Nobel Prize for Chemistry in 1972. The award was for their work on "understanding of the connection between chemical structure and catalytic activity of the ribonuclease molecule".

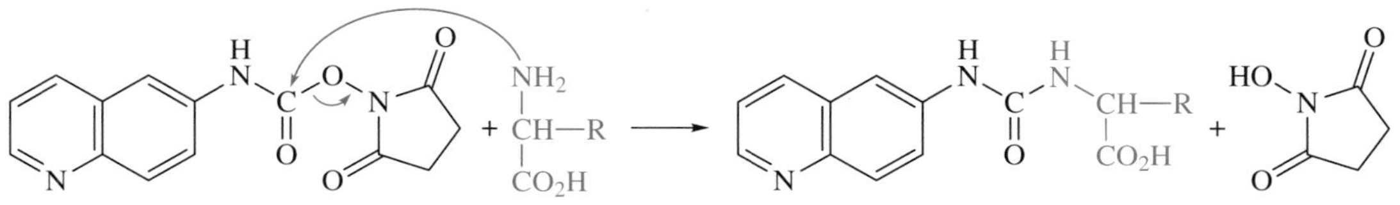

Scheme 4.3

4.4 N-terminal Analysis

It is important to have methods for identification of the N-terminal amino acid residue. For example, if a protein shows two different N-termini, then it usually contains two different polypeptide chains. Similarly, if a protein shown to be a dimer with chains of equal sizes shows only a single N-terminal residue, then it is added evidence that the chains are identical.

What is required is a label that can be attached to the amino terminal group, and which stays in place when the protein is hydrolysed. The first such label to be introduced was the **2,4-dinitrophenyl (DNP)** group. The N-terminus was labelled by the reaction shown in Scheme 4.4 and, after complete hydrolysis of the protein, the liberated N-terminal DNP-amino acid was identified by paper chromatography.

> There is another possibility. Sometimes proteins are partially degraded at the N-terminal ends, during purification, by **aminopeptidases**. This will lead to a population of molecules with different N-termini. The protein is said to have **ragged ends**.

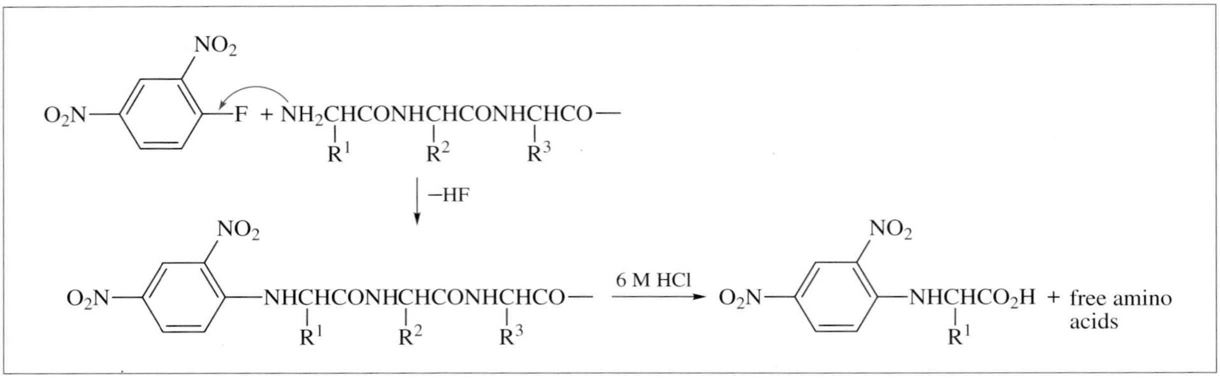

Scheme 4.4

This method is now largely of historic interest because it is relatively insensitive. A more modern method uses the reagent **5-(*N*,*N*-dimethyl-amino)naphthalene-1-sulfonyl chloride (Dansyl-Cl, DNS-Cl)**. The reaction of this reagent with a peptide, and hydrolysis to liberate the DNS-amino acid, is shown in Scheme 4.5. The advantage of this technique is that the DNS-amino acids are fluorescent. Identification is by separation of the products on **two-dimensional thin layer chromatography (2-D TLC)** followed by viewing under UV light. DNS-amino acids fluoresce yellow. The method is very sensitive and it is easy to identify 50 pmol of the product.

Scheme 4.5

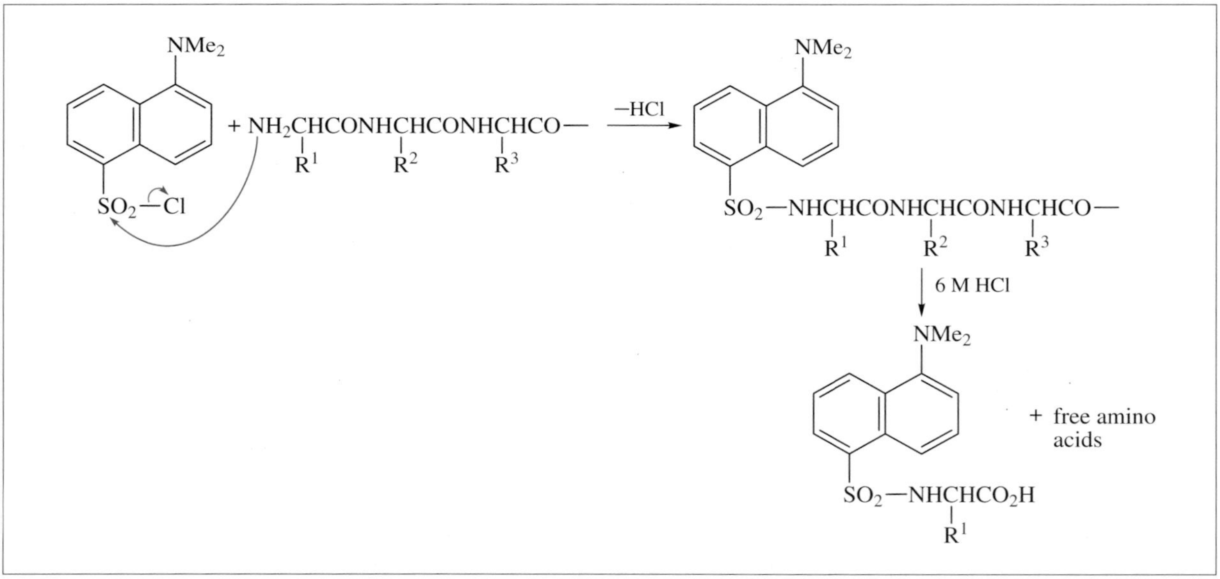

Worked Problem 4.3

Q A purified protein was shown to migrate as a single band on SDS-PAGE in the absence of mercaptoethanol. N-terminal analysis by the Dansyl method showed both glycine and leucine at equal intensities. Give possible explanations for these results and suggest an experiment that would give more information.

A The two N-terminal residues suggest that the protein has two different polypeptide chains. The fact that a single band was seen after SDS-PAGE indicates either that the two chains are identical in size or, more likely, that there are two different chains linked by disulfide bridges. To check this, SDS-PAGE should be carried out in the presence of mercaptoethanol. The two chains will now run separately and give rise to two different bands of lower M_r. The combined M_r values should be the same as that of the intact molecule, or be a simple fraction of it. For example, the protein might consist of four polypeptide chains, two of one sort and two of another.

In 2-D TLC the mixture to be analysed is spotted at one corner of a square thin layer plate. Chromatography in the first solvent separates the components along one edge of the plate. The plate is then dried, turned through 90°, and chromatography run in the second dimension.

4.5   Sequence Analysis by Chemical Methods

4.5.1   Peptide Sequencing

The first protein to have its amino acid sequence determined was insulin. This was done by Frederick Sanger in the late 1950s.[6] It was an enormously important piece of work, not least because it showed that proteins do indeed have defined amino acid sequences; this was still a matter of debate at the time. In essence, the two chains were separated and each broken into many small peptides. Each peptide was analysed for its amino acid composition, and the N-terminus determined using 1-fluoro-2,4-dinitrobenzene as described in Scheme 4.4. We will not go into details about how the sequences of the two chains were re-assembled from the results obtained because modern methods are somewhat different. Suffice it to say that the structure was successfully determined, and Sanger was awarded the Nobel Prize for Chemistry for this work in 1958.

What Sanger lacked was a method for degrading a polypeptide chain one residue at a time from the terminus while leaving the rest of the chain intact. A major breakthrough in peptide sequencing was made Pehr Edman, who developed such a method.[7] The procedure, now known as the **Edman degradation**, is shown in Scheme 4.6. The peptide or protein is reacted with **phenyl isothiocyanate (PITC)** in slightly alkaline solution. The unprotonated N-terminal amino group attacks the carbon of the

Sanger was awarded a second Nobel Prize for Chemistry in 1980 – a unique achievement. The second prize was for "contributions concerning the determination of base sequences in nucleic acids". So, remarkably, he made the seminal contributions to methods for determination of the structures of both of the major classes of biological macromolecules. He shared the Prize for 1980 with Walter Gilbert, who developed a different approach to sequencing nucleic acids.

isothiocyanate group and a **phenylthiocarbamyl (PTC)** peptide derivative is produced. The crucial reaction is the next one. If the PTC derivative is heated in anhydrous TFA, a ring-closure reaction occurs, which results in breakage of the peptide bond between residues 1 and 2. There are two essential points. Firstly, the rest of the peptide is left intact; hence it can be subjected to further rounds of degradation, each time liberating a new N-terminal amino acid. Secondly, the **thiazolinone** formed still contains the side chain of the amino acid so that, in principle, it can be identified, thus identifying the original N-terminal residue. The N-terminal residue has been coloured brown in Scheme 4.6 to make its fate clear.

Scheme 4.6

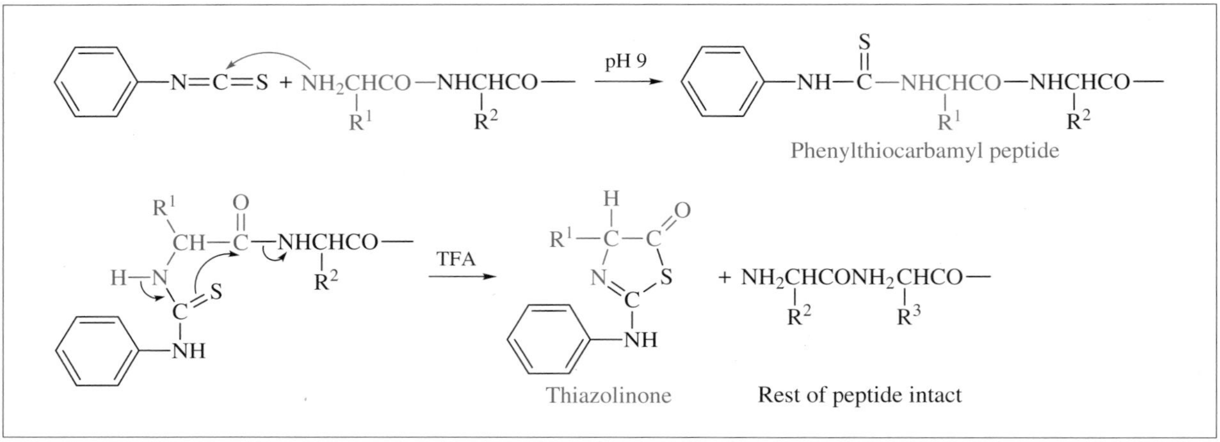

Phenylthiocarbamyl peptide

Thiazolinone   Rest of peptide intact

In fact, thiazolinones are not very stable, and are not suitable for identification. However, if the thiazolinone is heated in acidic solution, it undergoes rearrangement to the corresponding **phenylthiohydantoin (PTH)** derivative (Scheme 4.7). These are stable and can be readily identified. Originally this was done by thin layer chromatography, but the sensitivity of analysis has been enormously increased by using RP-HPLC on C_{18} reverse phases. An elution profile of a standard mixture of PTH-amino acids is shown in Figure 4.3. The detection limit is 10 pmol, so very small amounts of peptides or proteins can be analysed.

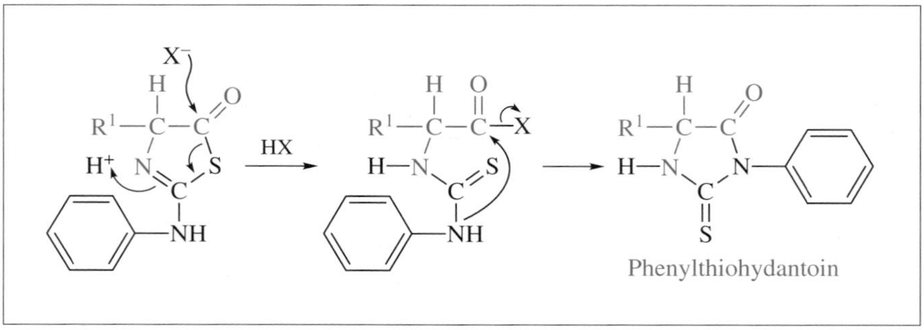

Phenylthiohydantoin

Scheme 4.7

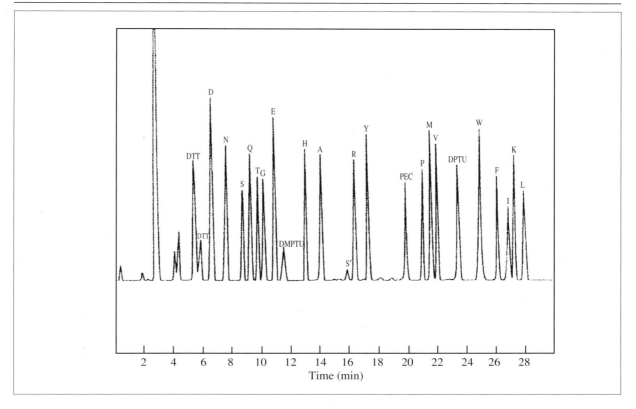

Figure 4.3 Separation of a standard mixture of PTH-amino acids by RP-HPLC. Peaks are identified by the usual one-letter codes. Other peaks are side products of the Edman reaction

Worked Problem 4.4

Q A peptide was found to have the amino acid composition Gly (1), Asp (2), Phe (1). It was subjected to four cycles of Edman degradation and the PTH-amino acid identified at each stage by RP-HPLC (refer to Figure 4.3). The elution times of the PTH-amino acids at each cycle were: cycle 1, 6.4 min; cycle 2, 25.9 min; cycle 3, 6.4 min; cycle 4, 10.0 min. What was the sequence of the peptide?

A Asp-Phe-Asp-Gly (or DFDG). This is one of 12 possible sequences consistent with the given amino acid composition.

The Edman degradation has been automated and, provided extremely pure reagents are used, it is capable of very high repetitive yields. This is important because, as we have seen before, the overall yield of a process decreases drastically if the step yield is low. It is particularly important that the cleavage step of the degradation is essentially complete, otherwise there will be "carry over" of a given residue into the next cycle. So, for example, if the first cleavage is only 95% complete, 5% of residue 1 will be seen as a contaminant at step 2. Another problem that arises is the "preview" of residues. If there is a peptide bond in the molecule that is partially labile under the reactions conditions of the degradation, then the amino acid at the C-terminal side of that bond will begin to appear before it is reached in the sequence. In short, there is a limit to the length of a peptide that can be sequenced by repeated application of the Edman degradation. It will vary from case to case, but is unlikely to be more than 40–50 residues. However, in favourable cases, the complete sequence of a peptide in this size range can be obtained by repeating the Edman degradation until the C-terminal residue is reached. Using these methods, both of the chains of insulin could be completely sequenced; so, at least in this context, insulin can be considered to be a large peptide.

Particularly with longer peptides, the question arises as to whether the sequence determined is correct. The results may be difficult to interpret towards the C-terminus because of decreasing yield, accumulating carry-over, and possible preview of residues. One check is that the sequence determined agrees with the amino acid composition. It may also be advisable to check the sequence at the C-terminus. There is no good chemical way of doing this, but Nature provides a useful tool. Enzymes called **carboxypeptidases** hydrolyse peptides and proteins at the C-terminal end and release free amino acids (Scheme 4.8). The best one to use is the enzyme from yeast, called carboxypeptidase Y, which has a wide specificity; that is, it removes any of the amino acids even though the rate of hydrolysis differs from one amino acid to another. Of course, once the C-terminal amino acid is liberated, the enzyme will proceed to remove the next one. For this reason it is necessary to do a **time course** where samples are removed at various time intervals and subjected to amino acid analysis. From the results it should be possible to obtain a short C-terminal sequence to confirm the results from Edman degradation.

Scheme 4.8

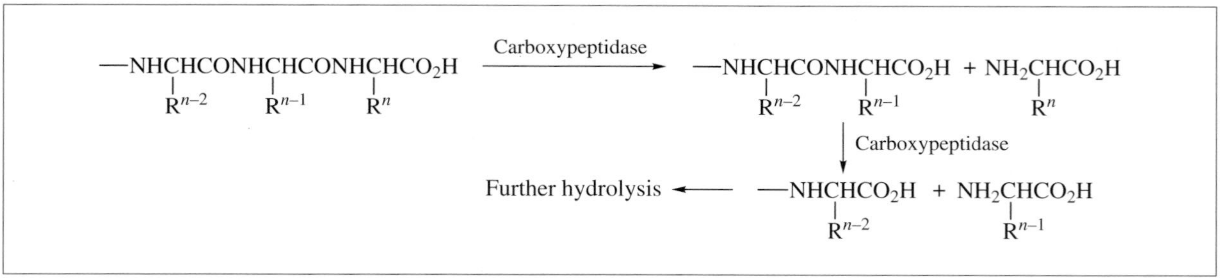

$$—NHCHCONHCHCONHCHCO_2H \xrightarrow{\text{Carboxypeptidase}} —NHCHCONHCHCO_2H + NH_2CHCO_2H$$

$$R^{n-2} \quad R^{n-1} \quad R^n \qquad\qquad\qquad R^{n-2} \quad R^{n-1} \qquad R^n$$

$$\downarrow \text{Carboxypeptidase}$$

$$\text{Further hydrolysis} \longleftarrow —NHCHCO_2H + NH_2CHCO_2H$$

$$R^{n-2} \qquad\qquad R^{n-1}$$

4.5.2 Problems with Cystine and Cysteine

Peptides and proteins containing cysteine and/or cystine present particular problems. As stated earlier, these residues are usually oxidized with performic acid before amino acid analysis. Suppose that such an analysis reveals four residues of cysteic acid. What was the situation in the original peptide? Were there four cysteines, or two cysteines and one disulfide bridge, or two disulfide bridges? The easiest way to answer this question is by reaction of the protein with **5,5′-dithiobis(2-nitrobenzoic acid)**, which is known as **Ellman's reagent**. This reacts with free thiol groups as shown in Scheme 4.9. The thiolate anion liberated is strongly coloured at alkaline pH, with a λ_{max} of 412 nm (that is, the solution becomes yellow). If a known amount of peptide is treated with Ellman's reagent, then the ΔA_{412} value allows calculation of the number of mol of cysteine (if any) per mol of peptide.

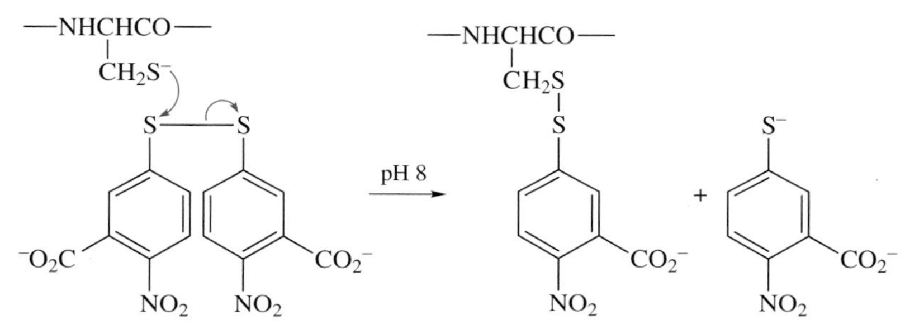

Scheme 4.9

Worked Problem 4.5

Q 0.100 mg of a peptide of M_r 2800 was dissolved in 3 mL of buffer in a cuvette of 1 cm path length. Excess Ellman's reagent was added and the final change in A_{412} observed was 0.307. How many cysteine residues does the peptide contain? Take the molar absorptivity of the thiolate anion as $\varepsilon = 13,600\ mol^{-1}\ dm^3\ cm^{-1}$.

A The concentration (c) of thiolate formed is obtained from $A = \varepsilon \times c$, since the path length of the cell was 1 cm. Hence $c = 0.307/13,600 = 2.25 \times 10^{-5}\ mol\ dm^{-3}$. However, the volume was 3 cm³, so the amount of thiolate formed was $2.25 \times 10^{-5} \times (3/1000) = 6.75 \times 10^{-8}$ mol or 67.5 nmol. The amount of peptide analysed was $0.1 \times 10^{-3}/2800 = 3.57 \times 10^{-8}$ mol or 35.7 nmol. Hence, within the limits of accuracy of the experiment, the peptide contained 2 mol of cysteine/mol.

For sequence analysis, it is usual to convert any cystines to cysteine (Scheme 4.1), and then to protect the sulfur by reaction with iodoacetic acid (iodoethanoic acid) to give carboxymethyl-cysteine (Scheme 4.10).

Sequence analysis can then proceed in the normal way. Note, however, that there is a problem if the original peptide contained two or more disulfide bridges. These were reduced and carboxymethylated before sequencing, so the sequence obtained contains no information about how they were joined in the first place (recall the discussion about apamin in Section 2.9).

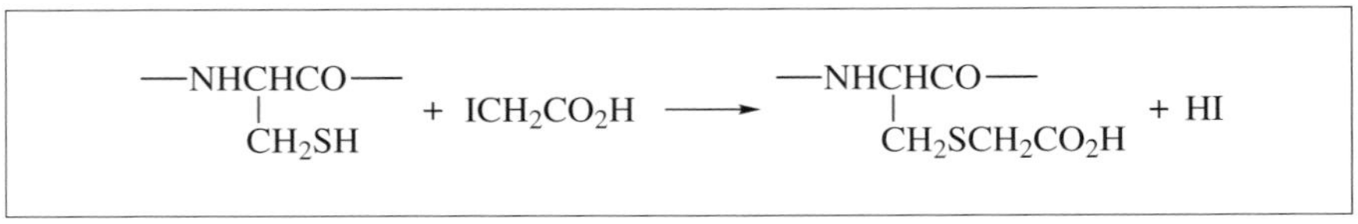

Scheme 4.10

For a solution to this problem we have to go back to the native peptide with bridges intact. The general approach is to partially hydrolyse the peptide (commonly using dilute sulfuric acid), and separate the product smaller peptides. It is hoped that one or more of these will have an intact bridge. The bridged peptide is then oxidized, the two halves separated, and analysed. With luck, this will provide the necessary information.

It is easier to see how this works with a specific example. In the case of apamin, the experiment was as summarized in Scheme 4.11. A disulfide-containing peptide was isolated and the two halves were found to have the amino acid compositions:

(a) $CysSO_3H$, Lys, Ala, Asp, Pro; and (b) $CysSO_3H$, Arg, Glu

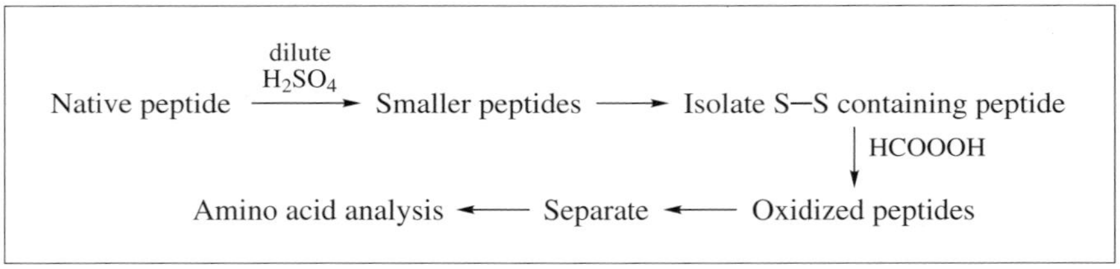

Scheme 4.11

From the amino acid sequence (**18** in Chapter 2), it is clear that peptide (a) contains cysteine at position 3, and peptide (b) contains cysteine at position 15 (remember that the cysteines have been oxidized to cysteic acid, and that asparagine and glutamine will appear as aspartic and glutamic acids on amino acid analysis). Hence one bridge links residues 3 and 15. The other, by elimination, links positions 1 and 11. In general, it is necessary to isolate and analyse $N - 1$ different peptides to locate N bridges.

There is a convenient way of identifying which peptides in the partial digest contain disulfide bridges by using **thin layer electrophoresis**. The mixture of peptides is spotted onto one corner of a square thin layer plate (usually coated with cellulose) moistened with a buffer of pH about

2. All of the peptides will be positively charged and so, on application of an electric field, will migrate towards the cathode to produce a line of spots along the edge of the plate (Figure 4.4A). If the plate is turned through 90° and electrophoresis repeated under the same conditions, then the peptides will migrate the same distance in the second dimension and will lie on a diagonal (see B). If, on the other hand, the plate is exposed to performic acid vapour before the second electrophoresis, then peptides containing disulfide bridges will be oxidized. Each such peptide will give rise to two product peptides, both of which now contain a negatively charged cysteic acid residue. On second-dimension electrophoresis these peptide pairs will lie off the diagonal; peptides not containing cystine will lie on the diagonal as before (see C). The pairs of peptides can be eluted from the thin layer plate and subjected to amino acid analysis. The method is known as **diagonal electrophoresis**.

Figure 4.4 Diagonal electrophoresis of disulfide-containing peptides. Diagram A shows the positions of the peptides after electrophoresis in the direction indicated; all peptides migrate towards the cathode at pH 2. In B, electrophoresis has been run under the same conditions in the second dimension and peptides lie on the diagonal. In C, the plate was exposed to performic acid vapour and then run in the second dimension. Peptides with disulfide bridges split into two and, because of the introduced negative charge, run more slowly off the diagonal (shown in brown)

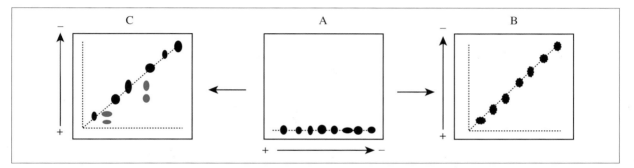

4.5.3 Extension to Sequencing Proteins

Clearly, a protein several hundred amino acids long cannot be completely sequenced by repeated application of the Edman degradation. We have already seen that a maximum of perhaps 40 or 50 residues can be identified in this way. The usual approach to any problem that is too large for a direct solution is to break it down into smaller pieces and solve them individually. This is what is done in protein sequencing. Suppose that we have a method for breaking the protein down into polypeptides at specific points in the chain. This is shown schematically in Scheme 4.12, where the protein is represented as a line along which there may be some hundreds of amino acids. Only the N-terminus and

Scheme 4.12

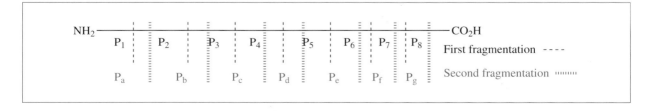

C-terminus are specifically shown. The protein is fragmented at the points indicated by the dashed black lines to produce a set of peptides, labelled P_1, P_2, P_3, and so on. These can be purified by methods such as RP-HPLC and sequenced independently. In the best case, all of the amino acids in the original protein will be included in this set of peptides.

The problem is that we do not know in what order the fragments occurred in the original protein. We can establish that P_1 was at the N-terminus by sequencing of the intact protein. What is not known is whether P_1 was followed by P_2, or by P_3, or any other of the fragments.

This difficulty can be solved by the **method of overlaps**. What is required is a second method for fragmenting the protein to produce a second set of fragments. The points of breakage in this case are shown in brown in Scheme 4.12. A new set of peptides, P_a, P_b, P_c, and so on, is generated. Again these are isolated and sequenced. The object is to find peptides in the second set that contain the C-terminal sequence of one of the peptides from the first set, and the N-terminal sequence of another. For example, in Scheme 4.12, peptide P_d contains the C-terminal region of peptide P_4 and the N-terminal region of peptide P_5. That is, peptide P_d overlaps peptides P_4 and P_5. In fact, the sequences of the second set of peptides in this idealized case contain all the information required to put the first set in order along the chain.

The success of the method of overlaps depends on the availability of procedures for fragmenting proteins at specific amino acid residues. In addition, to profit from the ability to sequence large fragments using the automated Edman method, the fragments should be large. There are two ways of doing this: using chemical methods and using enzymes.

Living organisms contain a large number of **proteolytic enzymes** or **proteinases** whose function is to hydrolyse proteins. They come in two types. **Exoproteinases** cleave amino acids from the end of the polypeptide chain (carboxypeptidases cleave at the C-terminus; aminopeptidases cleave at the N-terminus). We are interested here in proteinases that cleave in the interior of the protein. These are known as **endoproteinases**. To be useful they must be selective; many are not. For example, the enzyme **pepsin**, which is involved in digesting protein in the diet, has a very wide specificity and will cleave proteins on the C-terminal side of most aliphatic and aromatic amino acids. It thus produces many small peptides that are not useful for sequence studies. Much more useful is **trypsin**. This cleaves proteins at the C-terminal side of lysine and arginine residues – that is, the basic amino acids – to leave the lysine or arginine at the C-terminus of the product peptides (Scheme 4.13). Even this degree of specificity is less than optimal since lysine and arginine are relatively commonly occurring in proteins. The selectivity of the enzyme can be improved by treatment of the protein with maleic anhydride (*cis*-butenedioic anhydride) (Scheme 4.14). This converts the lysines to acidic

It is rarely as easy as this. For a variety of technical reasons that need not concern us, it is sometimes necessary to generate several sets of peptides to complete the sequencing project.

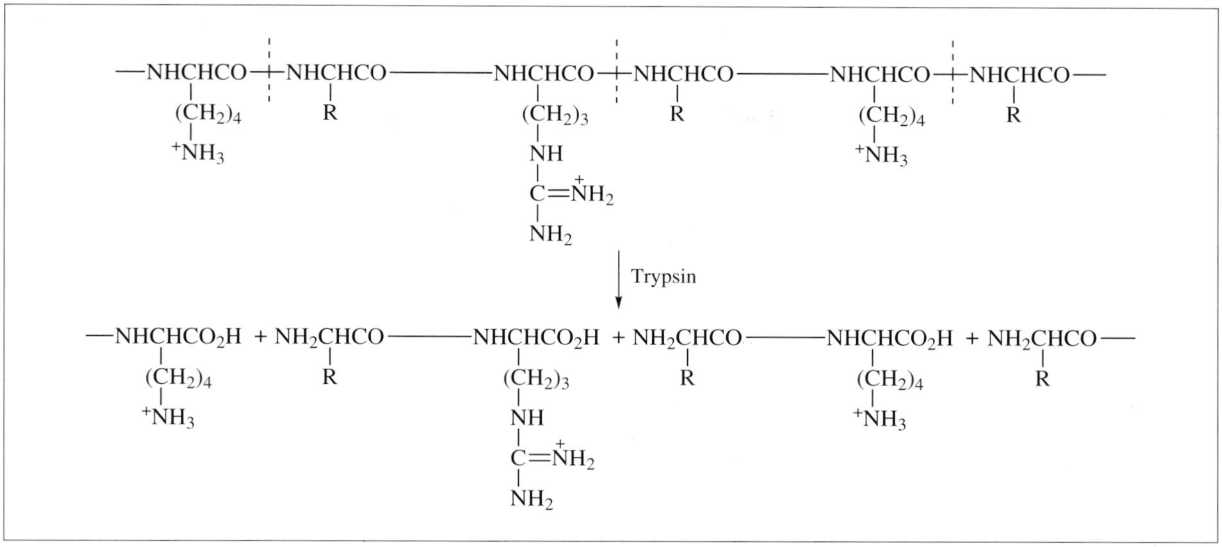

Scheme 4.13

derivatives that are not recognized by trypsin. The protein will now be cleaved only at arginines. Typically, a protein with 500 amino acids might contain 15–20 arginine residues, so trypsin cleavage will now produce a relatively small number of large peptides as required. There is also an enzyme called V_8 proteinase, isolated from the bacterium *Staphylococcus aureus*, which cleaves specifically at the C-terminal side of glutamic acid residues, and one from the fungus *Armillaria mellea*, which cleaves at the N-terminal side of lysine residues. These are useful complements to trypsin in sequence analysis. Thermolysin, which cleaves on the N-terminal side of aromatic and long-chain aliphatic residues, is also widely used.

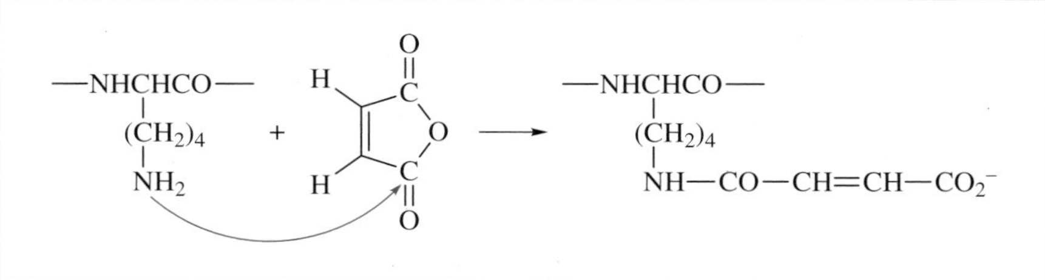

Scheme 4.14

Chemical methods are available for cleavage C-terminal to tryptophan, between asparagine and glycine, and C-terminal to methionine. The last of these is the most specific and, because methionine is one of the least-commonly occurring amino acids, it produces a small number of very large fragments. The reagent used is cyanogen bromide, and the reaction is shown in Scheme 4.15. Chain cleavage is accompanied by conversion of the C-terminal methionine to homoserine lactone.

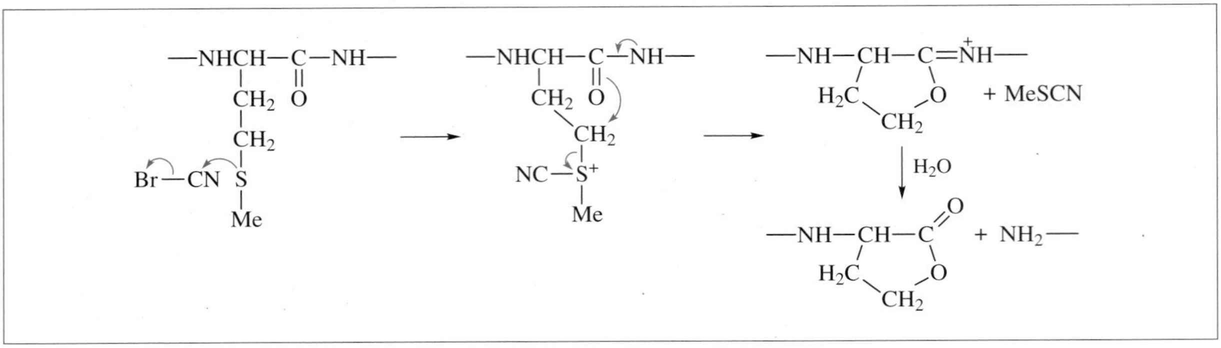

Scheme 4.15

There is also a method for cleavage N-terminal to cysteine residues. The reagent used is 2-nitro-5-thiocyanatobenzoic acid. This reacts with cysteine to form the *S*-cyano derivative which, under alkaline conditions, undergoes ring closure with peptide bond cleavage (Scheme 4.16). Unfortunately, this reaction leaves the N-terminus of the product peptide blocked, and so not available for Edman degradation. The N-terminal blocking group can be opened and converted to alanine by treatment with Raney nickel (which is hydrogen absorbed onto finely powdered nickel), but the reaction is not very efficient and is rarely used. Chemical cleavage at cysteine is, however, useful for generating fragments for analysis by mass spectrometry (see Section 4.6.2).

Scheme 4.16

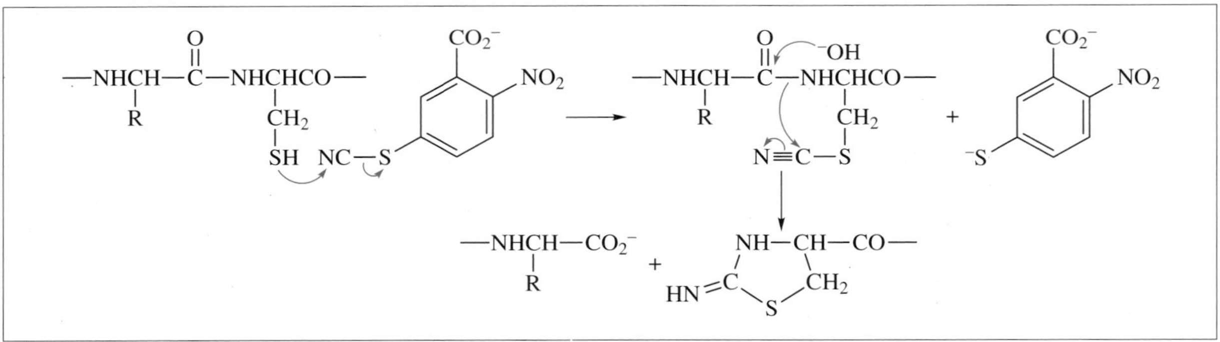

To summarize, a possible approach to sequence analysis of a protein would be:

1. Reduce and carboxymethylate to break any disulfide bridges and protect cysteines.
2. Fragment with CNBr, purify the product peptides, and sequence them using automated Edman degradation.
3. Maleylate the protein to block lysine side chains, digest with trypsin, isolate and sequence the product peptides.
4. Use peptides from the second digest to overlap peptides from the first set.

5. If the data are not complete, carry out another digest using, for example, V_8 proteinase.
6. If necessary, locate disulfide bridges using partial hydrolysis of the native peptide followed by diagonal electrophoresis.

Box 4.2 Protein Sequences by the Back Door

Protein sequencing using these classical methods is difficult, time consuming and requires a large amount of pure protein. Indeed, it is now relatively rare for complete sequences to be done this way. An alternative is to change the problem into one of sequencing a DNA molecule, which is, in fact, much easier. DNA molecules 400–500 bases long can be sequenced in a single operation and several such molecules can be handled at the same time. The methods of doing it are covered in any of the standard textbooks of biochemistry. Once a DNA sequence is obtained, then the corresponding protein sequence can simply be read from it based on the known **genetic code**. Each triplet of bases along the chain corresponds to a particular amino acid.

How is this done to solve a particular protein sequencing problem? For the genetic message to be expressed, the DNA is first **transcribed** into a related molecule, mRNA (the "m" stands for "messenger"; mRNA carries the genetic instructions from the nucleus to the site of protein synthesis in the cell). The mRNA is then **translated** into protein. So the route is

$$DNA \rightarrow mRNA \rightarrow Protein$$

It is possible to isolate total mRNA from a cell and convert it back into its **complementary DNA** (or **cDNA** as it is called) using the enzyme **reverse transcriptase** (this is an enzyme produced by certain viruses, called retroviruses, whose genomes are made of RNA rather than DNA; when the virus infects a host cell, the reverse transcriptase converts its RNA genome into DNA, and this DNA is inserted into the host's genome). This cDNA can be introduced into bacterial cells, and colonies of the cells grown up, each colony or **clone** originating from a single cell containing a unique inserted cDNA molecule. The trick is to find the clone containing the cDNA for the protein whose sequence we wish to know. To do this, we need to know a bit of the amino acid sequence, but this is easy enough using Edman degradation on the intact protein. Once a bit of sequence is known, an **oligonucleotide** can be synthesized which would code for a stretch of, say, 6–7 amino acid residues. [In fact, a mixture of oligonucleotides will be required since the genetic code is **degenerate**. Although two amino acids (Met and Trp) are coded by single triplets of bases or **codons**, the rest have more than one codon, and Ser and Arg are coded by as many as six different triplets. The mixture of oligonucleotides will have to cover all possible coding sequences. If possible, a stretch of sequence containing Trp and Met plus other amino acids with low degeneracy will be selected]. One component of the mixture of synthetic oligonucleotides will bind to the cDNA molecule coding the protein of interest, and can be used as probe to find the clone. Once the clone has been identified, the cDNA can be removed from it and sequenced. This will immediately lead to the amino acid sequence of the protein.

Even though this is obviously a lot of work, it may still be easier than directly sequencing the protein. More important, once the cDNA has been obtained, there are a lot of other interesting things that can be done with it. For example, specific mutations can be introduced into the cDNA which will result in single amino acid changes in the product protein. This technique is called **site-specific mutagenesis**. Studies on how the properties of the protein are affected by such changes give invaluable insights into structure/function relationships.

Sometimes much of the work has already been done. The complete genome sequences of several organisms have now been determined (see Chapter 6). Most of these are from bacteria, and in these cases it is easy to catalogue the amino acid sequences of all the proteins that the organism makes. It is then simply a case of identifying the protein of interest (see Chapter 6). With higher organisms, such as humans, it is more difficult. The reason is that the vast majority of protein coding genes in these species are not continuous stretches of DNA. Rather, the gene is split into coding sequences (called **exons**) separated by non-coding regions (called **introns**). It is not unusual for a gene to consist of a dozen or more exons, and for the amount of DNA in introns to exceed that in exons. The problem is that we do not have good methods for recognizing where exons end and introns begin, so it is not possible to directly translate genomic DNA sequences into protein sequences. When the mRNA for a particular protein is made, the non-coding regions of the gene are cut out. This is why cDNA is used for cloning experiments of the type described above, rather than genomic DNA.

This is a highly "potted" account of part of the fascinating fields of **molecular genetics** and **molecular cell biology**. They merit much more attention than can be given to them here. Standard texts on the subject are recommended under Further Reading.

4.6   Sequencing using Mass Spectrometry

Mass spectrometry has a distinguished history in the determination of the structures of organic molecules, but its application to the study of peptides and proteins is relatively recent. Early work was hampered by the fact that mass spectrometry was restricted to volatile compounds. This meant that proteins were inaccessible to the technique; to study peptides, they had to be chemically modified to make them volatile. Recent technical developments have removed these restrictions, and mass spectrometry has now become an extremely powerful tool in protein chemistry. These two phases in application of the method will be dealt with separately.

4.6.1 Early Approaches

Peptides are involatile because of the charged and polar residues that they contain. A solution to this problem was developed by Howard Morris and colleagues.[8] First, primary amino groups were converted to the methyl amides by treatment with acetic anhydride (Scheme 4.17). Then all hydrogen atoms on N, O and S were replaced by methyl groups by treatment with methyl iodide in the presence of the (methyl-sulfinyl)methyl carbanion (*e.g.* Scheme 4.18); the carbanion is generated by reaction of dimethyl sulfoxide with sodium hydride.

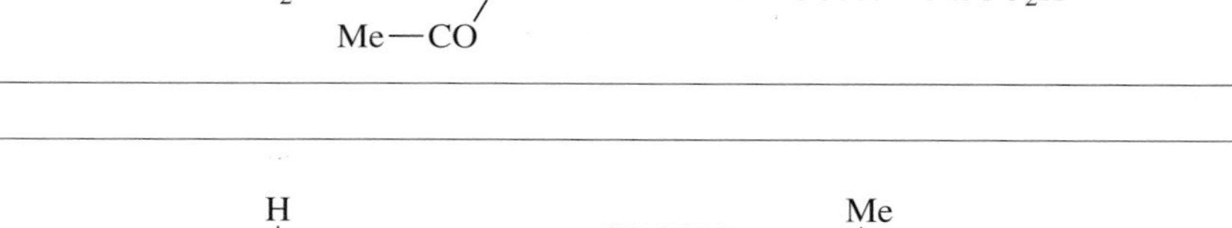

$$—NH_2 + \begin{matrix} Me—CO \\ \diagdown \\ O \\ \diagup \\ Me—CO \end{matrix} \longrightarrow —NH—COMe + MeCO_2H$$

Scheme 4.17

$$\begin{matrix} H \\ | \\ —N—CO— \end{matrix} + MeI \xrightarrow{\ ^-CH_2SOMe\ } \begin{matrix} Me \\ | \\ —N—CO— \end{matrix}$$

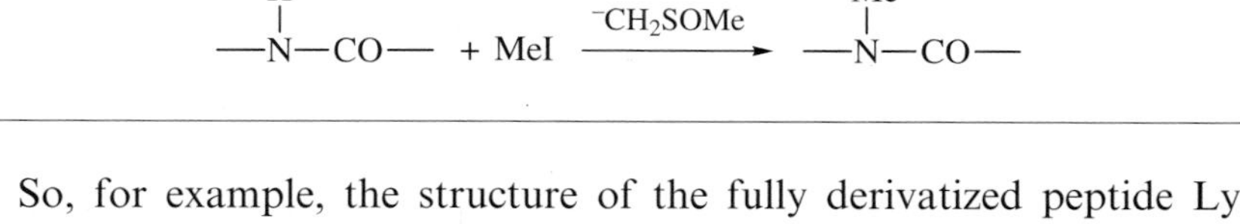

Scheme 4.18

So, for example, the structure of the fully derivatized peptide Lys-Asn-Asp would be as shown in **2** (atoms of the original peptide are in black and introduced atoms in brown).

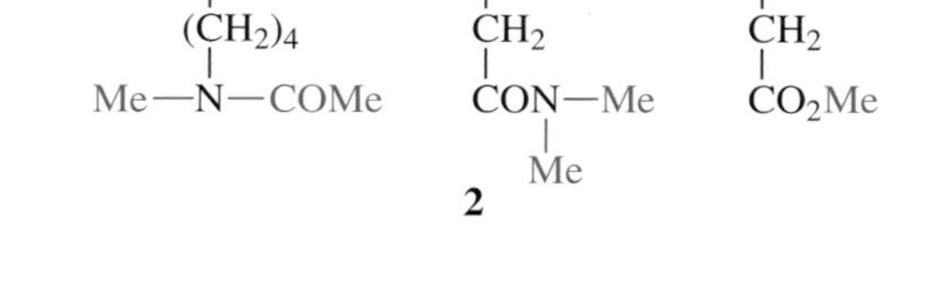

$$\begin{matrix}
& Me & & Me & & Me & \\
& | & & | & & | & \\
MeCO—N—CH—CO\!-\!\!&N—CH—CO\!-\!\!&N—CH—CO_2Me \\
& | & & | & & | & \\
& (CH_2)_4 & & CH_2 & & CH_2 & \\
& | & & | & & | & \\
Me—N—COMe & & CON—Me & & CO_2Me \\
& & & | & & \\
& & & Me &
\end{matrix}$$

2

If such a derivatized peptide is subjected to **electron impact mass spectrometry (EI-MS)** the parent ion will be seen but, more importantly, fragmentation will also occur. In particular, a set of fragments will be produced arising from breakage at the peptide bond; that is, at the points shown by dashed lines in **2**. So, in this case, the mass spectrum would appear as shown schematically in Figure 4.5. This is an idealized spectrum; there will be other peaks present arising from further fragmentation, but these have not been shown. The sequence of the peptide can be read directly from the mass spectrum, taking into account the relative molecular masses of the amino acids (given in Table 1.1) and allowing for the modifications carried out. For example, glycine as the N-terminal residue has an m/z of 114 (**3**), as an internal residue it has an m/z of 71 (**4**), and as the C-terminus it has an m/z of 102 (**5**). In the case shown in Figure 4.5, the peak at m/z 241 corresponds to N-terminal lysine. The m/z of the next peak is 156 units greater, which corresponds to an internal asparagine. The m/z of the final peak is 174 units greater again, which corresponds to C-terminal aspartic acid. Note that all of the amino acids give unique m/z values at each of the three possible positions except that, of course, it is not possible to distinguish between leucine and isoleucine, which are structural isomers.

Mass spectrometry measures the mass to charge ratio (m/z) of the ions. In EI-MS the ions are usually singly charged, so the m/z values in Figure 4.5 can be considered as masses.

$$\overset{\text{Me}}{\underset{|}{\text{MeCONCH}_2\text{CO}^+}}$$

3

$$\overset{\text{Me}}{\underset{|}{\text{NCH}_2\text{CO}^+}} \qquad \overset{\text{Me}}{\underset{|}{^+\text{NCH}_2\text{CO}_2\text{Me}}}$$

4   **5**

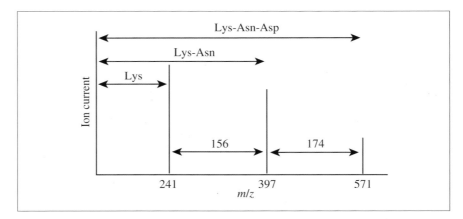

Figure 4.5 Electron impact mass spectrum of a derivatized tripeptide. The three-letter abbreviations are used here for the derivatized amino acids

This approach to peptide sequencing has some important merits. For example, it is possible in principle to sequence mixtures of peptides, provided that they differ in volatility. The peptides are introduced into the mass spectrometer from a heated probe, so that if the temperature of the probe is gradually raised, the spectra of the peptides will appear in turn. That of the most volatile peptide appears first. As the temperature is increased, this spectrum fades out, and that of the next most volatile peptide appears, and so on.

Perhaps the greatest advantage of the method is that it allows sequencing of peptides that have blocked terminal residues. Although not specifically referred to in Section 4.5.1, it should be obvious that the Edman degradation only works if the N-terminal residue has a free primary or secondary amino group. Similarly, carboxypeptidases require a free terminal carboxyl group. Many peptides are chemically modified at one end or the other, or sometimes both (see structure **17** in Chapter 1), and some proteins also have blocked N-terminal residues. In these cases, mass spectrometry provides a convenient way of sequencing the peptide and of identifying the blocking group.

Worked Problem 4.6

Q An N-terminal peptide was isolated after digestion of the α-chain of haemoglobin from trout blood with trypsin.[9] Its amino acid composition was found to be Ala (1), Leu (1), Lys (1), Ser (2). It was resistant to degradation by the Edman method. Consequently, it was derivatized using deuterated acetic anhydride [$(CD_3CO)_2O$] and MeI, and subjected to mass spectrometry. The resulting mass spectrum is shown in Figure 4.6; for clarity, only the sequence ions are shown. Deduce the amino acid sequence and identify the blocking group.

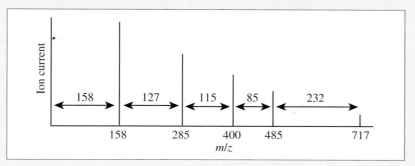

Figure 4.6 Mass spectrum of the N-terminal peptide from trout haemoglobin

A The C-terminus of the peptide would be expected to be lysine because of the specificity of trypsin. Consistently, the C-terminal m/z of 232 corresponds to lysine with three extra mass units arising from the fact that acetylation was done with deuterated acetic anhydride. The m/z of 85 corresponds to alanine, that of 115 to serine, and that of 127 to leucine. The m/z of 158 for the N-terminal residue corresponds to serine with an isotopically normal acetyl group. Acetylation was done using deuterated acetic anhydride, so the isotopically normal acetyl group must have been present in the original peptide. The sequence is, therefore, AcSer-Leu-Ser-Ala-Lys.

An interesting example of the power of mass spectrometry is shown by the discovery of a previously unknown derivative of an amino acid during sequence analysis. The N-terminal region of the protein prothrombin,

$$NH_2CHCO_2H$$
$$|$$
$$CH_2$$
$$|$$
$$CH$$
$$HO_2C \diagdown \quad \diagup CO_2H$$
6

which is involved in the process of blood clotting, yielded peptides which seemed to contain modified glutamic acid residues.[10] Analysis by mass spectrometry showed that the peptides contained γ-carboxyglutamic acid (**6**). This residue had gone undetected during sequence analysis using the Edman degradation because it decarboxylates during the process. In fact, prothrombin was found to contain 10 residues of γ-carboxyglutamic acid. Conversion of glutamic acid to the γ-carboxyl derivative occurs after synthesis of the protein (that is, it is a **post-translational modification**), and requires an enzyme that depends on vitamin K for its activity.

4.6.2 Modern Methods

Technical Advances

There has recently been an enormous increase in the applications of mass spectrometry to the study of peptides and proteins, arising from technical advances in the field (for a review, see Mann et al.[11]). The crucial advance has been the development of methods that allow analysis of unmodified peptides and proteins. Two methods are now available for transfer of these molecules into the gas phase, and for their ionization so that mass spectra can be recorded.

The first of these methods is **matrix-assisted laser desorption ionization** (referred to as **MALDI**). In this technique the protein is precipitated with a matrix material on to a metal support and allowed to dry; the most common matrices are α-cyano-4-hydroxycinnamic acid and dihydrobenzoic acid. The solid formed is then irradiated with nanosecond-long pulses from a nitrogen laser, which has a wavelength of 337 nm. The matrix absorbs the laser radiation and transfers some of the energy to the protein. This causes the protein to be desorbed from the matrix into the gas phase and to become ionized by capture of a proton. Hence the parent peak in the spectrum is that of the $[M + H]^+$ ion where M here represents the molecule being analysed.

The second method is **electrospray** (or **ES**). Here, a solution of the protein is pumped at very low flow rates (a few mm^3/min) through a hypodermic needle which is maintained at a high voltage. This causes fine droplets of the solution to be sprayed out. These rapidly evaporate and the vaporized protein molecules become charged in the process by gaining protons. In this case, the peptide or protein may gain more than one proton, so that multiple peaks are seen in the spectrum arising from $[M + H]^+$, $[M + 2H]^{2+}$, and so on. A particularly valuable technique exploiting ES is **liquid chromatography–mass spectrometry** (**LC–MS**) in which a mixture of peptides or proteins is separated by RP-HPLC and the effluent from the column directly analysed. In this way, the M_r values of the components of the mixture can be obtained.

The other area in which there have been major advances is in the instrumentation for the mass spectrometry itself. MALDI is usually carried out using **time-of-flight (ToF)** instruments in a technique known as **MALDI-ToF**. Here, the sample in the spectrometer is irradiated by a laser pulse to generate ions, and the ions are accelerated by an electrical field. They then enter a flight tube, and the time required for each ion to reach a detector is measured. The time taken depends on the m/z of the ion. The final spectrum is obtained by averaging signals from many laser pulses. ES is usually carried out using **quadrupole** mass spectrometers. A quadrupole consist of four rods to which an oscillating electrical field is applied. Depending on the amplitude of the field, only ions with a particular mass can pass through the quadrupole, and so a scan of the ions reaching a detector as a function of the amplitude of the field yields a mass spectrum. This type of MS is particularly useful for peptide sequencing as described below.

SELDI-ToF (surface-enhanced laser desorption ionization) is a variant of MALDI-ToF in which the sample to be analysed is absorbed onto a surface called a ProteinChip. The nature of the surface can be varied, depending on the properties of the proteins to be analysed. For example, a hydrophobic surface can be used to selectively absorb the hydrophobic proteins from a crude mixture.

Intact Peptides and Proteins

Mass spectrometry is now routinely used in the study of intact peptides and proteins. An example of the mass spectrum of a protein is shown in Figure 4.7. The M_r of the antibody was determined from the $[M + H]^+$ and the $[M + 2H]^+$ peaks as 150,364.3.

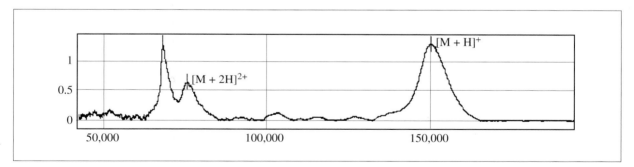

There are many applications of mass spectrometry of intact peptides and proteins.[12,13] For example, it is a good way of confirming the structure and purity of a synthetic peptide. Deviation of the measured M_r from that calculated from the sequence of the peptide can be correlated with problems such as failure to remove a protecting group, or failure of the coupling procedure at one step in the synthesis. It is also a very valuable method for analysis of post-translational modifications of a protein whose amino acid sequence has been determined indirectly from nucleic acid sequencing. For example, if the measured M_r is 42 units greater than that predicted from the sequence, this provides good evidence that the N-terminus is acetylated (an extra CH_3CO and loss of H). Many proteins after

Figure 4.7 SELDI-ToF mass spectrum of the antitoxin against *Clostridium botulinum* type A toxin absorbed on a normal phase (NP-1) ProteinChip. The spectrum shows both the [M + H]+ ion and the [M + 2H]2+ ion of the antibody. The other ion present was from a carrier protein (bovine serum albumin). (Spectrum courtesy of Dr Renata Culack)

synthesis are **phosphorylated**, usually on a serine residue, or less commonly on threonine or tyrosine. This introduces an extra 80 mass units into the protein. A particularly important application of mass spectrometry is the determination of the M_r values of proteins separated by two-dimensional electrophoresis. This technique is exploited in the new field of **proteomics** which is dealt with in Chapter 6.

Peptide Sequencing

This is done using **tandem mass spectrometry**, usually referred to as **MS-MS**. The protein is digested using one of the methods described in Section 4.5.3. The mixture of peptides is then introduced, by electrospray, into a quadrupole mass spectrometer. By adjustment of the amplitude of the quadrupole field, one of the peptide ions is selected and allowed to pass into the next section of the machine. Here the peptide is bombarded with atoms of a noble gas, such as neon or argon. This increases the vibrational energy of the peptide and causes it to fragment and form positive ions; this process is called **collision-induced dissociation (CID)**. The fragment ions are then passed into the final section of the mass spectrometer, which may be either a quadrupole or ToF, and the mass spectrum recorded. Another peptide can then be selected from the first stage of the mass spectrometer, fragmented and analysed, and so on.

Fragmentation of the peptide occurs mainly at the peptide bond. Two predominant sets of ions are produced, the origins of which are shown for a tetrapeptide in **7** (the H^+ outside the bracket indicates that there is a proton somewhere on the peptide). Ions of the y-type contain the intact C-terminus. The structures of y_1 and y_2 are shown in **8** and **9**. Ions of the b-type contain the N-terminus. The structures of b_1 and b_2 are shown in **10** and **11**.

These letters are used by convention to label the fragments shown in **7**. Other fragments (labelled a, c, x, z) are also produced but will not concern us here. A complete description of the fragments produced can be found in Allmaier.[12]

$$H_3\overset{+}{N}CHCO_2H$$
$$R^4$$
8

$$H_3\overset{+}{N}CHCONHCHCO_2H$$
$$R^3 \qquad R^4$$
9

$$H_2NCHC\equiv O^+$$
$$R^1$$
10

$$H_2NCHCONHCHC\equiv O^+$$
$$R^1 \qquad R^2$$
11

Interpretation of the mass spectra of these fragments is similar in principle to the method described in Section 4.6.1. Considering the y-type ions first, y_1 has the mass of the corresponding amino acid plus one; hence identification of y_1 identifies the C-terminal residue. The ion y_2 differs from y_1 by the atoms shown in brown in **9**, that is, by the next amino acid in the chain minus the elements of water. Consulting a list of the M_r values of the amino acids (Table 1.1) and subtracting 18 from each allows identification of this amino acid, and so on along the chain towards the N-terminus. Ions of the b-type provide a sequence starting from the N-terminal end. Ion b_1 has the mass of an amino acid minus 17. Ion b_2 is larger by the atoms shown in brown in **11**; that is, by the second residue in the chain minus the elements of water. Again, the process repeats until the C-terminal residue is reached.

Worked Problem 4.7

Q In the first stage of MS–MS, a peptide was found to show a protonated molecular ion $[M + H]^+$ at m/z 556. After CID it yielded a mass spectrum with ions at the following m/z values:

  b-type ions: 221, 278, 425, 538
  y-type ions: 132, 279, 336, 393
  What was its amino acid sequence?

A Consider the y-type ions. y_1 is the C-terminal residue plus a proton. Hence the C-terminal amino acid had $M_r = 131$. Table 1.1 shows that it must be either Leu or Ile. y_2 had a mass 147 Da greater than y_1, so the next amino acid had $M_r = 147 + 18 = 165$. This is Phe. y_3 had a mass 57 units greater, as did y_4. Hence both of the next two amino acids had $M_r = 57 + 18 = 75$. These are Gly. The parent ion had $M_r = 556$ which is 163 Da greater than y_4, so the N-terminal amino acid had $M_r = 163 + 18 = 181$. This is Tyr. Hence the sequence was either Tyr-Gly-Gly-Phe-Leu or Tyr-Gly-Gly-Phe-Ile. It is not possible to distinguish between these two possibilities. Amino acid analysis showed that the residue was Leu. The peptide was, in fact Leu-enkephalin (**3** in Chapter 2).

Note that m/z values are usually determined to the nearest whole number for sequencing purposes, and so to interpret the spectrum the M_r values in Table 1.1 should be rounded down.

Clearly it will not be a simple matter to identify the sequence ions in a complex spectrum which also contains ions arising from other fragmentation processes. There are, however, computer programs that can be used to help with the interpretation of the spectra. One way of simplifying the task of identifying sequence ions is to use trypsin to digest the protein. In this case all of the peptides, except that from the C-terminus, will end with either lysine or arginine. In these cases, the y_1 ion will have an m/z of either 147 (Lys) or 175 (Arg). Identification of one of these simplifies finding the y-series of ions. Note that lysine and glutamine have the same whole-number mass so cannot be distinguished at low resolution. In the case of a tryptic peptide, the C-terminal residue cannot be glutamine, so no confusion arises. The residues can also be distinguished if a high-resolution spectrum is recorded because they differ in mass by about 0.05 Da. Leucine and isoleucine cannot, of course, be distinguished.

Disulfide Bridges

Mass spectrometry provides a convenient way of establishing whether disulfide bridges are present in a peptide or protein, and also of locating bridges where two or more exist. Suppose, for example, that a peptide had been carboxymethylated, sequenced and found to contain two residues of CM-cysteine. Mass spectrometry of the native peptide will then establish whether the molecule contains two cysteine residues or a disulfide bridge, because the disulfide-bridged peptide would have an M_r two mass units lower than would the peptide containing free cysteines. Confirmation of the presence of a disulfide bridge could be obtained by reducing the peptide with mercaptoethanol (Scheme 4.1) and confirming that the M_r increases by two mass units.

   Location of the bridges where there are two or more is, in principle, straightforward once the sequence of the protein has been determined. The method involves digestion of the native protein with an enzyme such as trypsin and determining the m/z values of the product peptides. Scheme 4.19 shows a hypothetical example where there are two bridges. The protein is represented by a line, and the positions of the half-cystines, that is, the ends of the two bridges, are as shown. Also shown are the positions of lysines and arginines. Digestion of the protein with trypsin will yield a set of peptides P_1 to P_7 as indicated. P_1, P_4 and P_7 do not contain half-cystines and so will be formed as free peptides whose masses can be calculated from their known sequences. Peptide P_2, on the other hand, will be linked by a disulfide bridge to one of P_3, P_5 or P_6. Each of these possibilities would give rise to a bridged peptide whose mass can be calculated, and whose presence or absence in the mass spectrum of the mixture of ions can be established. So, for example, if the first and second half-cystines in the sequence are bridged, then a peptide ion will be observed with an m/z corresponding to P_2 plus P_3. A peptide should also be observed in this case with an m/z corresponding to P_5 plus P_6. In principle it should be possible to locate any number of bridges in a single experiment using this technique.

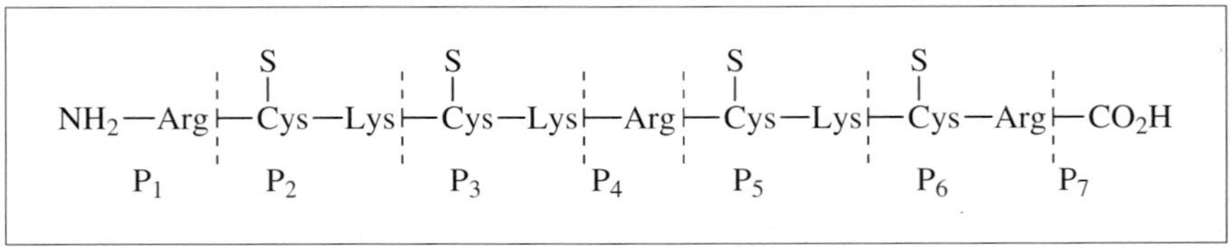

Scheme 4.19

Worked Problem 4.8

Q The protein recombinant human interferon α-2b, produced in genetically engineered *Escherichia coli* cells, has two disulfide bridges. The cysteines that constitue these bridges are in found in the following tryptic peptides:

CDLPQTHSLGSR (T$_1$)
ISLFSCLK (T$_5$)
FYTELYQQLNDLEACVIQGVGVTETPLMK (T$_{10}$)
YSPCAWEVVR (T$_{17}$)

The mass spectrum of the tryptic digest of the protein contained $[M + H]^+$ ions with *m/z* of 2118 and 4616.[14] After treatment of the protein with a reducing agent, these two peaks decreased in intensity and new ones appeared at *m/z* 911, 1210, 1314 and 3305. Deduce the positions of the disulfide bridges.

A The M_r values of the tryptic peptides can be determined from the M_r values of the constituent amino acids (using the values correct to 2 decimal places, remembering to subtract 18.01 for each peptide bond, and rounding the final values to whole numbers). These are T$_1$, 1313; T$_5$, 910; T$_{10}$, 3304; T$_{17}$, 1209. Linking a pair of these together by a disulfide bond will give an $[M + H]^+$ ion with *m/z* equal to the sum of the M_r values minus 2 for the disulfide bridge plus 1 for the proton. Combining T$_5$ and T$_{17}$ gives 2118. Combining T$_1$ and T$_{10}$ gives 4616. These are just the values observed in the spectrum so the disulfide bridges link the half-cystines in peptides T$_5$ and T$_{17}$, and T$_1$ and T$_{10}$. When the bridges are reduced, new peptides appear with $[M + H]^+$ ions at $M_r + 1$ as expected.

Summary of Key Points

- Empirical methods for determination of the relative molecular masses of proteins include gel permeation chromatography and SDS-PAGE.
- A protein has quaternary structure if it consists of two or more polypeptide chains that are not covalently linked. The chains, referred to as sub-units or monomers, may be the same or different.

- Comparison of M_r values obtained by gel permeation chromatography and by SDS-PAGE provides information about the quaternary structure, if any.
- There are two ways of determining amino acid compositions of peptides and proteins. One relies on separation of the amino acids by ion exchange chromatography followed by reaction with ninhydrin. The other involves conversion of the amino acids to fluorescent derivatives, followed by separation by RP-HPLC.
- The amino acid sequence of a peptide can be determined by sequential removal of amino acids from the N-terminus using the Edman degradation. The product phenylthiohydantoin derivatives are identified by RP-HPLC.
- Sequence analysis of proteins can be done by hydrolysis of the polypeptide chain at specific points, separation of the product peptides, and sequencing the individual peptides using the Edman degradation. The order in which the peptides occurred in the original protein is then established using the method of overlaps.
- The positions of disulfide bridges in the native protein are determined by hydrolysis of the molecule, isolation of peptides containing intact bridges, and analysis of their amino acid compositions.
- Electron impact mass spectrometry of peptides requires volatile derivatives to be made. For this reason it was only rarely used. The method did, however, achieve some popularity as a way of sequencing peptides with blocked terminal residues.
- Recent advances in mass spectrometry, particularly development of MALDI and ES methods, have made it possible to analyse intact, unmodified peptides and proteins. One important application is for determination of post-translational modifications.
- Tandem mass spectrometry (MS–MS) provides a powerful method for sequence analysis of mixtures of peptides and, hence, of proteins.

Problems

4.1. Figure 4.8 shows the results of SDS-PAGE for a set of standard proteins (marked with arrows in the left-hand lane) and purified aspartate aminotransferase (right-hand lane). The M_r values of the standards, reading from the top of the gel, were 45,000, 36,000, 29,000, 24,000, 20,000 and 14,000. Use these values to determine the M_r of aspartate aminotransferase. Comment on the choice of standards used in this experiment.

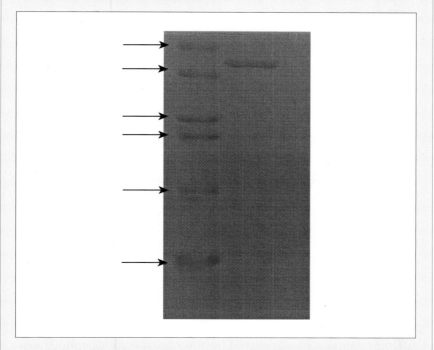

Figure 4.8 SDS-PAGE of a set of standard proteins (left-hand land) and purified aspartate aminotransferase (right-hand lane)

4.2. The cAMP-dependant protein kinase from human testis was found to have an M_r of 166,000, as determined by gel permeation chromatography. SDS-PAGE in the absence of mercaptoethanol showed two bands, one with an M_r of 43,000 and the other with an M_r of 40,000. N-terminal analysis on the intact protein showed both glycine and methionine. Provide an interpretation of these results.

4.3. The sedimentation coefficient S of haemoglobin at 20 °C is 4.41×10^{-13} s and the diffusion coefficient D is 6.30×10^{-11} m^2 s^{-1}. The partial specific volume $\bar{v} = 0.75$ cm^3 g^{-1}. Calculate the molar mass, assuming that the density of the buffer used was 1.01 g cm^3. Note that the T in equation (4.1) is the absolute temperature, and that the value of $R = 8.314$ J K^{-1} mol^{-1}.

4.4. A peptide was incubated with carboxypeptidase A. Every 5 min a sample was taken and analysed for liberated amino acids, with the results shown in Table 4.2. Use these results to determine as much as possible of the C-terminal sequence of the peptide.

Table 4.2 Amino acids liberated from a peptide by digestion with carboxypeptidase A.

Time (min)	Mol of amino acid liberated per mol of peptide			
	Gly	Leu	Phe	Trp
5	0.05	0.45	0.25	0.05
10	0.12	0.75	0.55	0.11
15	0.18	0.95	0.70	0.18
20	0.23	1.00	0.80	0.21

4.5. During sequence analysis of a protein using chemical methods the following three peptides were among those obtained after digestion of the protein with trypsin: ILSMR; TMADR; SELR. Cleavage with cyanogen bromide yielded the peptide ADRILSM. Digestion with thermolysin yielded the peptide LSMRSE. Use these data to assemble a partial sequence of the protein.

4.6. Refer to Worked Problem 4.6 on the sequence analysis of the N-terminus of the haemoglobin from trout. Trout blood contains a second species of haemoglobin whose α-chain has a somewhat different amino acid sequence. The N-terminal tryptic peptide from this second component was analysed in precisely the same way as before. The sequence ions observed in the spectrum were at m/z 158, 285, 414, 499 and 731. What is the difference in sequence between the peptides from the two proteins?

4.7. Refer to Worked Problem 4.7 on sequence analysis of Leu-enkephalin by mass spectrometry. Analyse the b-type sequence ions to confirm the sequence obtained. Note that, in this experiment, ion b_1 was not observed.

4.8. A peptide was synthesized by SPPS using the t-BOC strategy. After release from the resin and deprotection, a sample of the peptide was subjected to MALDI-ToF MS. Two major $[M + H]^+$ ions were observed. One had m/z expected for the target peptide. The other was 100 mass units greater. Give an explanation of this result.

References

1. J. M. Walker (ed.), *The Protein Protocols Handbook*, Humana Press, Totowa, NJ, 1996.
2. T. Svedberg and K. O. Pedersen, *The Ultracentrifuge*, Oxford University Press, Oxford, 1940.
3. B. E. C. Banks, S. Doonan, M. Flogel, P. B. Porter, C. A. Vernon, J. M. Walker, T. H. Crouch, J. F. Halsey, E. Chiancone and P. Fasella, *Eur. J. Biochem.*, 1976, **71**, 469.
4. D. H. Spackman, W. H. Stein and S. Moore, *Anal. Chem.*, 1958, **30**, 1190.
5. S. Cohen, K. DeAntonis and D. P. Michaud, in *Techniques in Protein Chemistry IV*, ed. R. Angeletti, Academic Press, San Diego, 1993, p. 289.
6. F. Sanger and H. Tuppy, *Biochem. J.*, 1961, **49**, 463.
7. P. Edman, *Acta Chem. Scand.*, 1950, **4**, 277.
8. H. R. Morris, D. H. Williams and R. P. Ambler, *Biochem. J.*, 1971, **125**, 189.
9. S. Doonan, A. G. Loudon, D. Barra, F. Bossa and M. Brunori, *FEBS Lett.*, 1978, **85**, 141.
10. H. R. Morris and A. Dell, *Biochem. J.*, 1976, **153**, 663.
11. M. Mann, R. C. Hendrickson and A. Pandey, *Annu. Rev. Biochem.*, 2001, **70**, 437.
12. G. Allmaier, in *Encyclopedia of Analytical Science*, ed. R. Townshend, P. J. Worsford, S. J. Haswell, R. Macrea, H. W. Werner and I. D. Wilson, Academic Press, London, 1995, p. 3003.
13. K. Biemann and H. A. Scoble, *Science*, 1987, **237**, 992.
14. B. N. Pramanik, A. Tsarbopolous, J. E. Labdon, P. P. Trotta and T. L. Nagabhushan, *J. Chromatogr.*, 1991, **562**, 377.

Further Reading

T. E. Creighton, *Proteins: Structures and Molecular Properties*, 2nd edn., Freeman, New York, 1993.

B. Levin, *Genes VII*, Oxford University Press, Oxford, 2000.

H. Lodish, A. Berk, L. Zipursky, P. Matsudaria, D. Baltimore and T. Darnell, *Molecular Cell Biology*, 4th edn., Freeman, New York, 1999.

5

Proteins in Three Dimensions

Aims

By the end of this chapter you should understand:

- The various ways in which models of protein structures can be drawn
- That the conformation of a peptide can be described in terms of two dihedral angles
- The characteristics of the major elements of secondary structure in a protein: α-helix, β-strand and β-turn
- The factors that determine the three-dimensional structures adopted by a protein
- Why proteins fold, and how the stability of the folded state can be determined
- How selected proteins carry out their biological functions

5.1   Introduction: the Anatomy of Crambin

A vast amount of effort has been expended on determination of the primary structures of proteins, and information of great biological interest has been derived from the results. For example, comparison of the sequences of the same protein from different organisms has lead to increased understanding of the processes of evolution. This topic will be returned to in Chapter 6. If, however, the question is asked how knowledge of the primary structure of a particular protein helps to understand how it works, the answer is that it does not. The reason for this is that proteins fold up in space into precisely determined three-dimensional structures, and it is on these three-dimensional structures that the activities of proteins depend. This chapter is concerned with what those structures look like and why they form. A small number of examples will also be used to show how activity and three-dimensional structures are related.

As an introduction to the subject we will look at the general features of the three dimensional structure of a small protein called crambin.[1] This protein comes from the seeds of the Abyssinian cabbage (*Crambe abbyssinica*). It has only 46 amino acid residues but still contains most of the essential structural features found in much larger proteins. Its primary structure is:

TTCCPSIVAR SNFNVCRLPG TPEAICATYT GCIIIPGATC
PGDYAN

It has three disulfide bridges linking residues 3→40, 4→32 and 16→26. Its three-dimensional structure is shown in a variety of different ways in Figure 5.1, always with the molecule in the same orientation.

All of the structural models in this chapter were produced using a graphics program called **RasMol**. The molecular coordinates required to produce the models were obtained from the database of the **Research Collaboratory for Structural Bioinformatics**; the database identification codes for proteins used as examples here are given in the text. For example, crambin is 1CRN. See Chapter 6 for a detailed discussion of these topics.

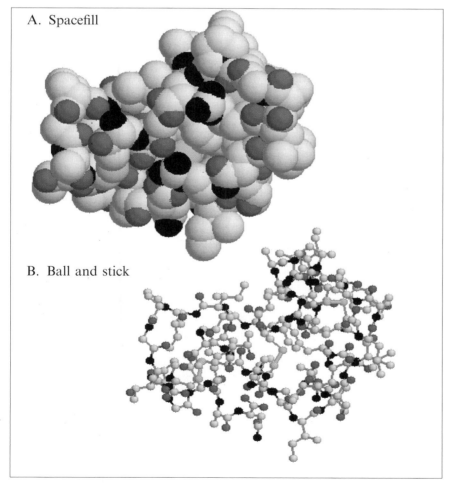

A. Spacefill

B. Ball and stick

Figure 5.1 Models of crambin

Figure 5.1 Continued

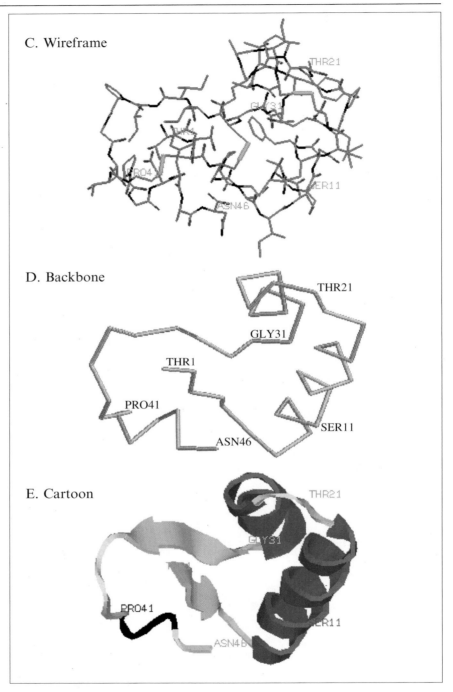

C. Wireframe

D. Backbone

E. Cartoon

Figure 5.1A shows a **spacefill** model in which each atom is represented as a sphere of radius proportional to its van der Waals radius. Note that, for clarity, hydrogen atoms are not shown in this or any other of the models; all spare valencies in the model are occupied by hydrogen. The colour code used for the atoms in this model is light grey for carbon, black for nitrogen, dark brown for oxygen and light brown for sulfur.

One thing that is apparent from this model is that the protein is folded up into a compact, roughly spherical, three-dimensional structure. This overall fold is known as the **tertiary structure**. The amino acids in the protein are in close contact and fill most of the available space. A space-fill model is useful to get an impression of the overall shape of the molecule, but it is difficult to see details of the structure, and impossible, of course, to see what is going on in the interior. Figure 5.1B shows a **ball and stick** model. Here, heavy atoms are drawn as balls of the appropriate colour, and the bonds linking them are drawn as sticks. The disulfide bonds are shown as light brown cylinders. It is now possible to see all of the residues and to identify some of them. So, for example, the residue at the bottom and slightly to right of centre in Figure 5.1B is an isoleucine [side chain $-CH(Me)CH_2Me$; remember that only the carbon atoms of the side chain are shown].

The ball and stick model is still very cluttered, and this is even more true when models are drawn of proteins containing several hundred amino acid residues. Better is the **wireframe** model in Figure 5.1C. Here, each line represents the bond joining a pair of atoms, but the atoms themselves are not shown. Where the bond joins two atoms of different types, the line is shown half in the colour of one atom and half in the colour of the other. This representation is less cluttered and it is possible, but not easy, to trace the polypeptide chain through the molecule. To help in this, every tenth amino acid and the C-terminal residue has been labelled with its number and name. It is instructive to try to trace the chain through the model, and to identify individual amino acid residues from the sequence of the protein given above.

To the unpractised eye, the structure in Figure 5.1C still seems to be without any obvious regularity or order. That this is not so is shown by the model in Figure 5.1D. This is a **backbone** structure where all the atoms have been removed and all that is shown is a line joining each of the C_α atoms of the polypeptide chain. It is now easy to trace the chain through the molecule. More importantly, regions become apparent where the chain seems to have a regular structure. For example, from position 7 to 17 the chain appears to be following a helical path. This part of the structure is, in fact, arranged in what is known as an α-helix, which is an example of one type of **secondary structure**. Similarly, from position 32 to 35 the chain is extended and is in a conformation known as a **β-strand**, the second main type of secondary structure. There is a third element of secondary structure in the molecule. This is a **β-turn** (sometimes referred to as a **β-hairpin**) where the chain undergoes an 180° change in direction in a sequence of four residues. The

This is not the colour scheme conventionally used by RasMol, where nitrogen is blue and sulfur is yellow. The colours have been changed here because blue and yellow are not available.

orientation of the molecule in Figure 5.1D does not allow this to be clearly seen.

These elements of secondary structure are of prime importance in describing and classifying three-dimensional structures of proteins. For this reason, a **cartoon** representation is often used to make them more obvious. Crambin is shown as a cartoon in Figure 5.1E. Here, the α-helices (there are two of them) are shown as dark brown spirals. The β-strands are shown as flat light brown ribbons with an arrowhead indicating the direction of the chain (this is useful in large molecules with many regions of β-structure). Again, there are two of them, and they lie alongside one another; the significance of this will be discussed in the next section. Finally, the single β-turn in the molecule is shown as a black region of the backbone. The rest of the protein, shown in grey, has no ordered secondary structure and is said to be in a **random coil** configuration.

In summary, then, proteins contain local regions of regular structure – α-helices, β-strands, β-turns – separated by segments of random coil. The local ordered regions constitute the secondary structure. The whole chain is folded into a tightly packed globular structure referred to as the tertiary structure. In the next section we will look at these conformations more systematically.

Box 5.1 Determination of Three-dimensional Structures

The vast majority of protein structures have been determined by **single-crystal X-ray diffraction** methods. The first diffraction pattern of a globular protein was obtained in the 1930s by Bernal and Hodgkin, but the first three-dimensional structure of a protein, that of the oxygen-binding protein myoglobin, was not obtained until 1966 by Kendrew. Shortly afterwards the structure of haemoglobin was determined by Perutz. The long time that elapsed before the first structures were obtained reflects the formidable problems that had to be solved in protein crystallography. What these problems are, and how they were solved, cannot be gone into here, but for an introduction to the subject the book by Rhodes[2] is hard to beat.

A few points do, however, need to be made. The first is that, clearly, crystallography cannot be done without crystals. Obtaining crystals of proteins is often the limiting factor in three-dimensional structure studies. Indeed, some proteins defy crystallization, and so their structures are inaccessible by this technique. This is particularly true of proteins that are component parts of biological

Dorothy Hodgkin did a great deal of important early work on the structures of biological molecules such as vitamin B_{12}. She was awarded the Nobel Prize in Chemistry in 1964 for "her determinations by X-ray techniques of the structures of important biological substances". She completed the determination of the structure of insulin in 1969. Max Perutz and John Kendrew were jointly awarded the Nobel Prize in chemistry in 1962 for "their studies of the structures of globular proteins".

membranes, and the structures of only a few such proteins have been determined to date.

A second important point is that protein crystallography does not identify the positions of individual atoms in protein crystals. What is determined is the **electron density** throughout the unit cell of the crystal, and it is rarely possible to do this at a **resolution** of better than about 2.5 Å (0.25 nm) because of the quality of the crystals; that is, features in the crystal of a size less than this cannot be individually distinguished. A resolution of the order of 1 Å would be required to "see" individual atoms. The solution of this difficulty requires knowledge of the amino acid sequence of the protein. The electron density map will contain features that are characteristic of the shapes of the side chains of particular residues, and so by comparing the map with the known sequence, the crystallographer seeks to fit the sequence into the electron density to produce a **model** of the structure. There follows a process of refinement in which the model is continuously improved, but it remains a model rather than a structure, and theories about, for example, how a particular protein works based on such models must always be viewed with caution.

There is another problem. Proteins carry out their functions in solution (usually). How do we know that the structures that exist in crystals are the same as those in solution? Perhaps the environment in the crystal causes the protein to change its shape. Fortunately, at least for small proteins, there is direct evidence that this is not so. It is now possible to determine the structures of proteins in solution using **multi-dimensional NMR spectroscopy**. Obviously, NMR spectra of proteins will be immensely complex, and their interpretation correspondingly difficult, but considerable success has been achieved in recent years (see Evans[3] for a review). The important point is that, where proteins have been examined both by NMR and by crystallographic methods, the structures obtained are essentially the same. This provides confidence that crystal structures are not an artefact of the environment of the protein in the crystal lattice.

5.2 Conformations of the Polypeptide Backbone

Each residue in a protein contributes three bonds to the polypeptide backbone, and so it might be thought that to describe the conformation of the chain requires specification of three **dihedral angles** for each residue. In fact this is not so, because the peptide bond itself is planar.

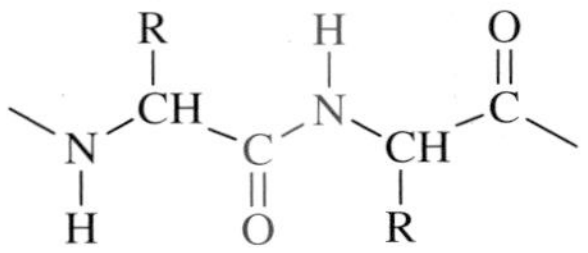

1

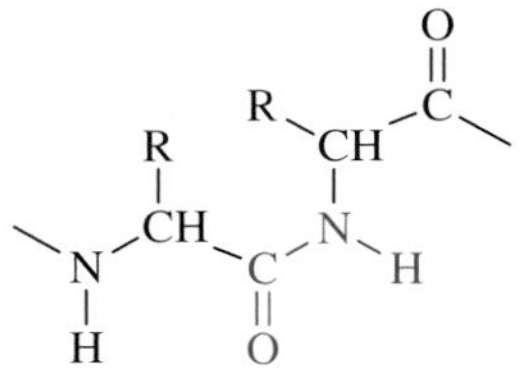

2

This is because it is a resonance hybrid between the structures shown in Scheme 5.1, and so the NH–CO bond is partially double in character. In principle, the peptide bond could have either a *cis* or a *trans* configuration, but *trans* is strongly favoured over *cis* (by a factor of about 1000:1) for steric reasons. Structure **1** shows a *trans* peptide bond, and **2** shows the corresponding *cis* form. It is clear that in the *cis* form there would be substantial steric interaction between the side chains on adjacent α-carbon atoms, and so this form is not favoured. The exception is with proline, where there is only a small difference in stability between the *trans* and *cis* forms. Consistently, some proteins have been found to contain one or more *cis* proline residues. This is not to say that a fraction of the molecules contain a given proline residue in the *cis* conformation with the rest *trans*; a particular proline will be either all *trans* or all *cis*.

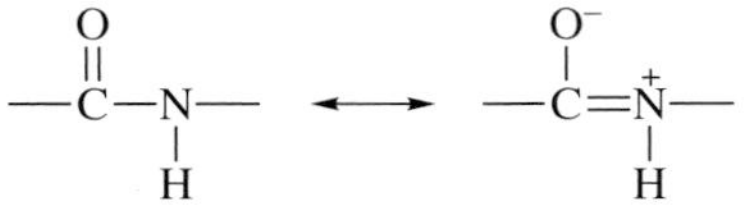

Scheme 5.1

Worked Problem 5.1

Q Draw the structures of a dipeptide with proline on the C-terminal side in both the *trans* and the *cis* configurations. Comment on their relative stabilities.

A The *trans* and *cis* configurations are shown in **3** and **4**, respectively. The *trans* configuration is marginally the more stable because in the *cis* form there will be steric interactions between the carboxylic acid group of proline and the side chain of the N-terminal residue (note that there is little double bond character in a peptide bond to proline, so the dihedral angle can change to relieve any steric interactions).

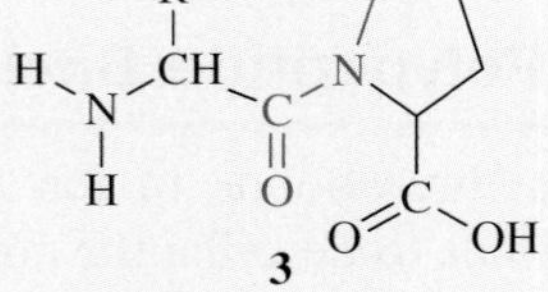

3

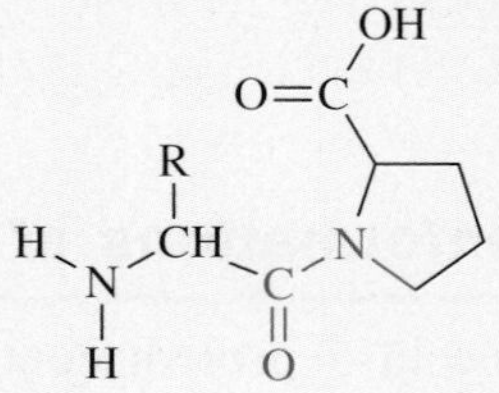

4

That leaves the other two bonds. Figure 5.2 shows the C_α atom of an amino acid with the preceding and following peptide bonds. The boxes contain the two planar, *trans* peptide units. Rotation is allowed around the N–C_α bond, and this dihedral angle is called ϕ (Greek letter phi). Rotation is also allowed around the C_α–C bond, and this dihedral angle is called ψ (Greek letter psi). Rotation is not allowed around the CO–NH (peptide) bond; this dihedral angle is called ω (Greek letter omega) and has a value of 180° for the *trans* peptide bond as shown. In the diagram in Figure 5.2 the other two dihedral angles have also been set at 180°, which results in a fully extended peptide chain. In principle, to define the conformation of a polypeptide chain amounts to defining the two dihedral angles ϕ and ψ for each residue in the chain.

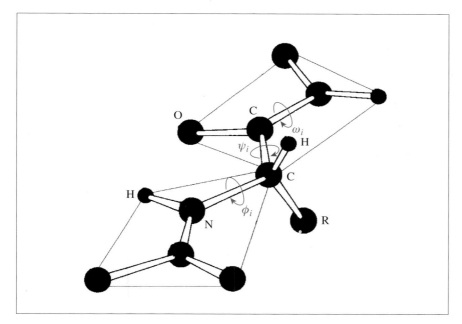

Figure 5.2 Dihedral angles of a peptide

It turns out that not all values of ϕ and ψ are possible, for steric reasons. Many combinations will result in unfavourable interactions between main chain atoms or side chains. This problem was analysed by Ramachandran and Sasisekharan,[4] who used solid sphere models of amino acids to determine the range of possible values of ϕ and ψ. The results for peptides made from L-amino acids are shown in Figure 5.3 as what is known as a **Ramachandran plot**. The fully allowed regions are shown enclosed by solid lines in Figure 5.3 and contain a relatively small area of the plot. The dashed lines enclose an area where ϕ and ψ are allowed with small deviations of the structure from ideal bond lengths and angles. Notice that the dashed region disappears off the plot at the top and reappears at the bottom because a dihedral angle of 180° is the same as one of –180°. In practice, the vast majority of residues in actual proteins have dihedral angles in the area enclosed by the dashed lines. The exceptions are usually glycine residues, which, because they have no side chains, are subject to fewer steric restrictions.

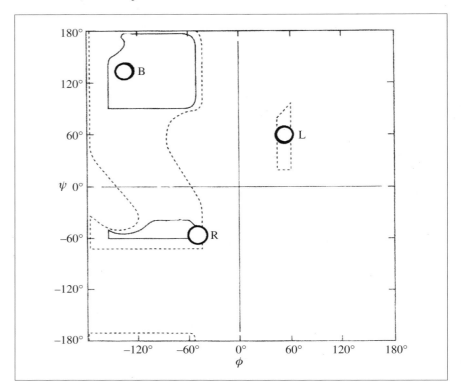

Figure 5.3 Ramachandran plot showing allowed combinations of dihedral angles in peptides. The circled regions show the range of dihedral angles in β-structures (B), right-handed α-helices (R) and left-handed α-helices (L) (Adapted from Ramachandran and Sasisekharan[4])

Worked Problem 5.2

Q Suppose that we wish to predict the conformation of a peptide with 11 amino acid residues. Assuming that each of the 10 peptide bonds can have dihedral angles in the regions marked B or R in Figure 5.3, how many possible conformations must we consider?

A Since there are two possibilities for each peptide bond, then the number of possible conformations is 2^{10}, or 1024. So even if the possible values of the dihedral angles are restricted to these small regions of the Ramachandran plot, the number of conformations to consider is very large. This is not a viable approach to the prediction of three-dimensional structures of peptides, and even less of proteins. See Chapter 6 for an alternative.

5.3   Regular Secondary Structures

5.3.1   The α-Helix

The secondary structure elements in crambin have already been outlined above, but it is necessary to look at them in more detail. The α-helix spanning residues 7 to 17 in crambin has been cut out and is shown separately in Figure 5.4. In Figure 5.4A, a wireframe model of the helix is shown with Ile-7 at the bottom. The polypeptide chain winds upwards and makes a complete turn every 3.6 residues. The essential feature is that the carbonyl oxygen of each amino acid makes a hydrogen bond with the hydrogen on the amide nitrogen of the fourth residue along the chain; the hydrogen bonds are shown as dotted lines between the oxygen and nitrogen atoms (recall that the hydrogen atoms are not explicitly shown in the models). It is these hydrogen bonds that give the α-helix its stability. Because the helix makes a turn every 3.6 residues, and the hydrogen bond is to the fourth residue along, this means that the hydrogen bonds are nearly parallel to the axis of the helix. The same view of the helix is given as a ball and stick model in Figure 5.4B. The helix is right handed; that is, as the helix is traced from the N-terminal end to the C-terminal end the direction of rotation is right handed, or clockwise. Figure 5.4C shows the helix looking almost along the axis from Ile-7 and with the C_α backbone drawn in; it is now easier to see the direction in which the helix turns. With a right-handed helix made from L-amino acids the side chains point outwards, away from the helix axis. The dihedral angles for a perfectly regular right-handed α-helix are $\phi = -57°$ and $\psi = -47°$. There is, however, some degree of latitude in these angles, as shown by the region marked in Figure 5.3; α-helices in proteins rarely have the ideal dihedral angles.

The existence of both the α-helix and the β-sheet in proteins was originally proposed by Linus Pauling based on studies of the structures of small molecules. Pauling was awarded the Nobel Prize in Chemistry in 1954 "for his research into the nature of the chemical bond and its application to the elucidation of the structure of complex substances". He was also awarded the Nobel Peace Prize in 1962.

The α-helix is by far the most common helical structure in globular proteins. It is not, however, the only one possible. The 3_{10}-helix has three residues per turn and hydrogen bonds from residue i to $i + 3$. The packing is very tight, and this type of helix is only observed at the ends of α-helices.

Box 5.2   Hydrogen Bonds

When hydrogen is bonded to an electronegative atom, the bond is polarized so that there is a partial positive charge on the hydrogen. Because hydrogen has no non-bonding electrons, the resulting partial positive charge interacts strongly with other electronegative atoms, and the resulting interaction is referred to as a **hydrogen bond**. The hydrogen bond can be considered as electrostatic in nature, and is usually represented as in **5**, where nitrogen has been chosen as the donor atom and oxygen as the acceptor. It is relatively weak, with bond energies in the range 10–30 kJ mol^{-1} (that is, about 10% of the value for a normal covalent bond). It is also

$$\overset{\delta-}{—N}\overset{\delta+}{—H}\text{- - -}\overset{\delta-}{O}—$$
5

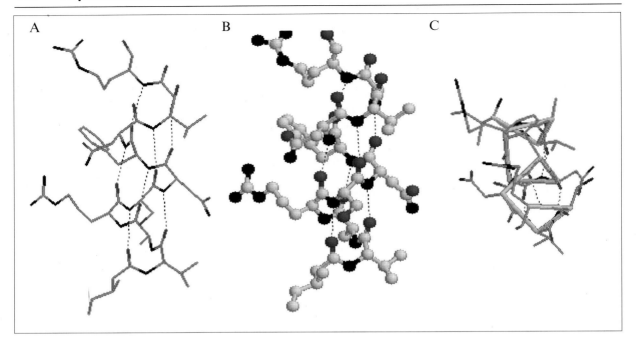

A   B   C

Figure 5.4 An α-helix from crambin. C is an end-on view showing the right-handed direction of the helix

a rather long bond. For example, in the case shown in **5** the H–O distance will be about 0.2 nm; that is, the N–O distance will be about 0.3 nm. In a protein structure, whenever a hydrogen bond donor and an acceptor atom are separated by about this distance, it is assumed that a hydrogen bond exists between them. A given hydrogen bond donor can interact with more than one acceptor; an example of this is shown in Figure 5.6A.

Hydrogen bonds are of enormous importance for the structures and activities of biological molecules. We are concerned here with the role that they play in the structures of proteins. It is, however, worth noting that it is hydrogen bonding that is responsible for the interactions between the two strands of a DNA molecule, and hence for the way in which genetic information is transmitted from parents to progeny.

5.3.2 β-Sheets

Note that Cys-4 and Cys-32 are linked by a disulfide bridge which joins the end of strand 1 and the beginning of strand 2. Do not confuse this with a continuation of the polypeptide backbone.

Residues 1–4 and 32–35 in crambin are in β-strands; that is, they have dihedral angles in the region indicated with a B in Figure 5.3. Figure 5.5A shows these two strands. They run in opposite directions and are hydrogen bonded to one another; they constitute a small **antiparallel β-sheet**. Generally, β-strands do not occur singly in proteins because they cannot form stabilizing internal hydrogen bonds. Instead, they occur in sheets where two or more strands lie alongside one another and form hydrogen bonding networks between the strands. There is an important difference here compared with the α-helix. In the latter, the hydrogen bonding is between amino acids close together in the sequence of the protein. In the case of β-sheets, the strands forming the sheet can come from widely separated parts of the protein, as, indeed, is the case with crambin.

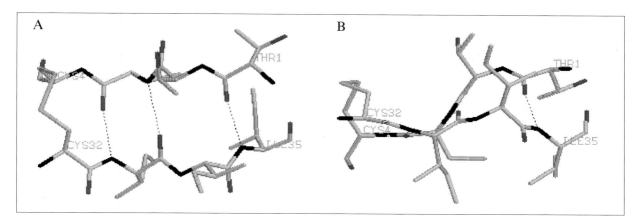

The chains in β-sheets are generally not fully extended but are puckered; this can be seen in Figure 5.5B where the sheet is viewed edge-on. Moreover, when several chains are associated together there tends to be a twist in the resulting β-sheet. In Figure 5.6 a β-sheet consisting of four antiparallel β-strands has been extracted from the protein prealbumin (1BMZ).[5] Figure 5.6A shows the β-sheet from the top. Note how the two internal strands make hydrogen bonds with their neighbours. Successive peptide units make hydrogen bonds first with a chain on one side, then with the chain on the other. Note also that at some positions a given group makes hydrogen bonds with two others rather than one. Figure 5.6B shows the β-sheet drawn as a cartoon and looking at it side-on. The puckering of each strand is obvious, as is the fact that the sheet twists from front to back.

Figure 5.5 The β-sheet from crambin. A: view from the top. B: view from the side

Note that, in Figure 5.6B, RasMol has interrupted the β-strand at the back with a short section of random coil. This is because of substantial deviations of the dihedral angles from ideal values.

Worked Problem 5.3

Q Draw three antiparallel peptide chains, each six residues long, in a β-sheet, and show the hydrogen bonding interactions between them. For simplicity, omit the side chains of the residues.

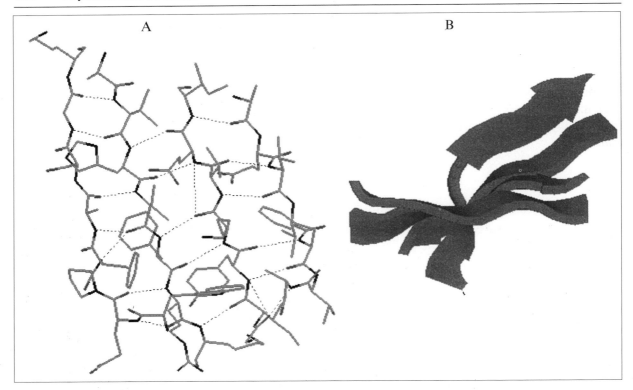

Figure 5.6 Antiparallel β-sheets. A: a four-stranded antiparallel β-sheet from prealbumin. B: the same β-sheet as a cartoon and viewed from the side

A The structure is shown in **6**, with arrows indicating the directions of the chains.

Strands can also run parallel to one another to form a **parallel β-sheet**, and sometimes a sheet contains a mixture of parallel and antiparallel strands. The enzyme triose phosphate isomerase is a dimer of identical subunits. Each subunit has a central core consisting of eight parallel β-strands which form a cylindrical structure known as a **β-barrel**. Three of the parallel strands have been extracted from the structure of the enzyme (1YPI)[6] in Figure 5.7A.

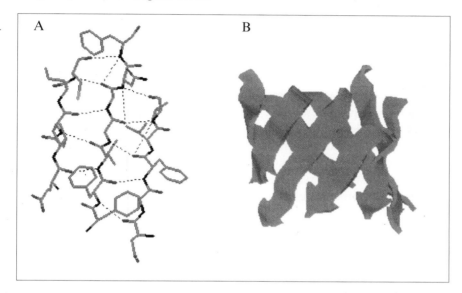

Figure 5.7 Parallel β-sheets. A: a parallel β-sheet from triose phosphate isomerase. B: the β-barrel from triose phosphate isomerase shown as a cartoon

Notice that in the parallel β-sheet the hydrogen bonds are more inclined to the direction of the chain than they are in the antiparallel sheet. Figure 5.7B shows a carton of the complete β-barrel; all the β-strands run from top to bottom in the figure. In the intact enzyme the β-barrel is surrounded by α-helices. Figure 5.8 shows a cartoon of

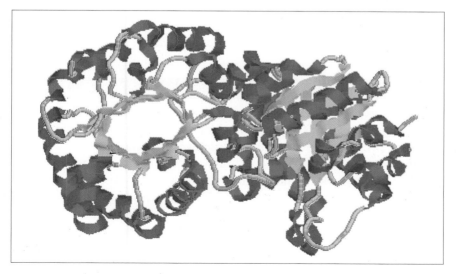

Figure 5.8 Cartoon of the dimeric structure of triose phosphate isomerase

the dimeric enzyme looking down the β-barrel of the left-hand monomer and sideways at the right-hand monomer. Recall that the two monomers constitute the quaternary structure of the enzyme.

5.3.3 Hairpin Bends

The remaining element of the structure of crambin that we have not looked at in detail is the β-turn. This is shown as a wireframe model in Figure 5.9. The essential point here is that the carbonyl group of the proline residue at the bottom of the figure is hydrogen bonded to the nitrogen of the tyrosine reside three positions along in the chain. The turn includes a glycine residue. This is very commonly the case because the lack of a side chain in glycine allows it to adopt the required, somewhat distorted, backbone angles required to fit into a β-turn. There are, in fact, several distinct types of β-turn that differ in the dihedral angles adopted by the central residues, but the details need not concern us here. There is also another type of turn, called a **γ-turn**, in which the first amino acid is hydrogen bonded to the residue two along in the chain. This is a very tight turn and occurs relatively rarely.

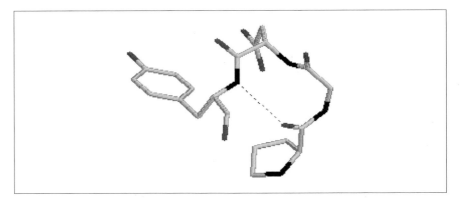

Figure 5.9 The β-turn from crambin

Box 5.3 Fibrous Proteins

Fibrous proteins provide some of the important structural elements in living organisms. They generally consist of extended polypeptide chains with regular secondary structures associated together to form fibres. **Silk fibroin**, for example, is the main component of webs made by spiders and of the cocoons made by silk worms. Its main constituents are glycine (40%), alanine (26%) and serine (12%). It contains polypeptide chains in antiparallel β-sheets stacked on top of one another to form the fibre. **Keratin**, the main component of hair, cuticle and feathers, is α-helical. Pairs of helices coil together to forrm **coiled coil** dimers, somewhat like two strands in a rope, and these then pack together to produce filaments that are about 10 nm in diameter. Similar coiled coils occur in the protein **myosin**, which is an important component of muscle.

A different sort of structure is found in **collagen**. Collagen is the major structural component of bone, tendon, ligaments, skin and blood vessels, and is the most abundant protein in animals. It has an unusual amino acid sequence in that it is largely composed of repeats of the tripeptide sequence Gly-X-Y with many of the X and Y residues being proline or hydroxyproline (**7**); the hydroxyproline residues are synthesized as a post-translational modification. This unusual composition is matched by a most unusual three-dimensional structure. Each chain forms a left-handed helix with 3.3 residues per turn and a rise per residue along the helix of 2.9 Å. This gives a rise of 9.6 Å for each turn of the helix compared with 5.4 Å for an α-helix. Three such helices coil around a central axis to form a right-handed super helix, as shown in Figure 5.10A. Every third residue is close to the axis of the super helix, where there is no space for a side chain; hence the requirement for glycine at this position. A short section of a single chain is shown in Figure 5.10B, from which the positions of the glycine residues can be seen (remember that no hydrogen atoms are shown). The structure in Figure 5.10 is not that of natural collagen, but rather of a synthetic peptide which has the characteristic Gly-X-Y repeating sequence and which forms a collagen triple helix (1BKV).[7]

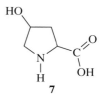

7

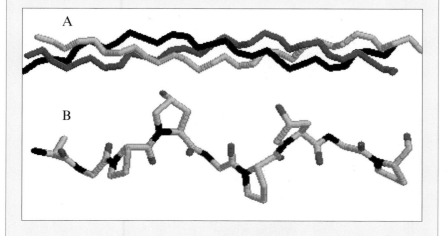

Figure 5.10 Structure of collagen. A: triple helical structure. B: wireframe model of a small section of one chain

5.4   Factors Determining the Three-dimensional Structures of Proteins

The picture that emerges from the discussion above is that proteins consist of regions of regular secondary structure folded into an overall globular shape. The secondary structure elements are formed largely as a result of their hydrogen bonding interactions, but what gives rise to the overall folding of the chain into a precisely defined tertiary structure? Part of the answer lies in the formation of β-sheets which, as we have seen, often bring together remote regions of the protein chain. In addition, however, each amino acid residue makes a variety of interactions with its neighbours in space. These are of several different types. Firstly, hydrogen bonds will be formed between the side chains of polar residues with other such residues, or with the peptide bond. There may also be ionic interactions between amino acids with ionized acidic and basic side chains. Because of its resonance hybrid character, the peptide bond itself is a dipole, and so can interact electrostatically with other dipoles or with charged amino acids.

Any given amino acid may or may not be involved in interactions of the types listed above. All of them, however, make van der Waals interactions with their neighbours. These interactions, though individually weak, are all-pervasive and collectively make a very important contribution to the shape of a protein. In Figure 5.11, one amino acid residue of crambin, Phe-13, was selected and all atoms within a distance of 4 Å (0.4 nm) from the benzene ring of its side chain drawn in spacefill mode.

Crystallographers almost always use the Ångström (abbreviation Å) as a unit of length. The unit is named after the 19th century Swedish physicist A. J. Ångström. 1 Å is equal to 10^{-10} m or 0.1 nm. It is a convenient unit because bond lengths are in the range 1–2 Å and so its use avoids decimals. It is, however, frowned upon by strict adherents of SI units!

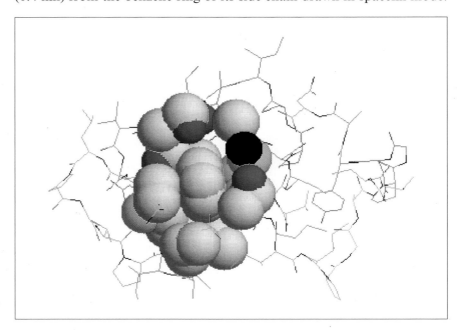

Figure 5.11 Packing around the side chain of Phe-13 in crambin

The benzene ring is viewed edge-on in the centre and can be seen to be in close contact with a variety of atoms. Interactions of this sort exist throughout the molecule, and adjustment of the folding so as to optimize van der Waals contacts is of central importance in determining its precise shape.

5.4.1 Why Proteins Fold Up

Proteins are synthesized as a linear array of amino acids and there is no machinery in the cell that causes them to fold. It was shown by Anfinsen[8] that protein folding is a spontaneous process, and that the ability to achieve the native fold is in some way encoded in the amino acid sequence. What he did was to take the enzyme **ribonuclease**, reduce its disulfide bridges, and then, by treatment of the enzyme with the denaturing agent urea, completely unfold it. The unfolded enzyme was, of course, completely inactive. He then slowly removed the urea, re-oxidized the protein, and showed that the native three-dimensional structure was regained.

Christian Anfinsen was awarded the Nobel Prize in Chemistry in 1972 for his "work on ribonuclease, especially concerning the connection between the amino acid sequence and the biologically active conformation".

The fact that folding is spontaneous means that the equilibrium in equation (5.1):

$$U \rightleftharpoons N \qquad (5.1)$$

where U stands for the unfolded protein and N stands for the native form, lies to the right. In other words, folding is associated with a decrease in free energy. The free energy change ΔG is related to the enthalpy change ΔH and the entropy change ΔS by equation (5.2):

$$\Delta G = \Delta H - T\Delta S \qquad (5.2)$$

where T is the absolute temperature. ΔH will approximate to the difference between the energies of interaction of the amino acid residues with one another in the native state, on the one hand, and the energy of interaction of those same amino acids with water when the protein is unfolded. The precise values of these cannot be calculated, but it is likely that interaction with water is the more favourable so that ΔH in equation (5.2) will be positive; that is, it will act against spontaneous folding. This means that the ΔS value must be positive if folding is to be spontaneous.

To see the significance of this, it is helpful to split the entropy term into contributions from the protein itself (ΔS_p) and contributions from the solvent (water) in which it is dissolved (ΔS_w). ΔS_p is certain to be negative. The reason for this is that in the unfolded state the protein chain can adopt many conformations; that is, it is a disordered system with high entropy. On folding, the protein is constrained to adopt a unique conformation, and so the process must be associated with a decrease in entropy. Indeed, by making some assumptions about the number of conformations that the unfolded chain can adopt, it is possible to make a rough estimate of the entropy change on folding. If there are n possible conformations per residue in the unfolded state, then ΔS_p is given by equation (5.3):

$$\Delta S_p = -NR \ln n \tag{5.3}$$

where N is the number of residues in the protein. The value of ΔS_p must necessarily be negative, and so for folding to be spontaneous, ΔS_w must be large and positive (*i.e.* $-T\Delta S_w$ large and negative). That this is so arises from the hydrophobic effect (Section 1.3.3). When the protein is unfolded, the hydrophobic amino acid side chains are exposed to the solvent. Water responds to this by forming ordered layers around these non-polar residues (see Section 1.3.3). When the protein folds, the hydrophobic side chains are hidden away in the interior of the molecule, out of contact with the solvent, and the constraints on the water are removed. This results in an increase in the entropy of the water which more than outweighs the loss of configurational entropy of the protein. Hence protein folding is essentially driven by the gain in entropy of the solvent, and has little to do with the interactions that the residues make within the folded molecule. This is not to say that those interactions are not important. The entropy effect causes the protein to fold whereas the intramolecular interactions dictate the precise shape into which it folds.

Worked Problem 5.4

Q Calculate, using equation (5.3), the change in entropy on folding for a protein of 100 residues, assuming that $n = 6$. Take R as 8.314 J mol^{-1} K^{-1}. Calculate $-T\Delta S_p$ at 25 °C (298 K).

A $\Delta S_p = -100 \times 8.314 \times \ln 6 = -1490$ J mol^{-1} K^{-1}. Hence $-T\Delta S_p = 444$ kJ mol^{-1} K^{-1}.

5.5   Stability of the Folded State and Ways of Measuring It

The three-dimensional structures of proteins are only marginally stable. Changes in temperature, or composition of the solvent, usually result in loss of the native structure and transition to an unfolded state. This process can be studied by measuring any property of the protein that changes as a result of the transition. For example, the **viscosity** of protein solutions increases when the protein is unfolded. However, most methods that are used for studies of this type involve spectroscopic methods (for a review of the application of these methods to proteins, see Sheehan[9]). The UV absorption spectra of proteins show a peak at around 280 nm, largely arising from their content of tryptophan and tyrosine. When proteins are unfolded, these amino acids move from the less polar interior of the protein into contact with the solvent, resulting in a decrease in both the λ_{max} and the absorption coefficient. The changes are, however, relatively small. More dramatic changes occur in the fluorescence spectrum. In aqueous solution, tryptophan fluoresces relatively weakly with an emission maximum at about 350 nm. In non-polar solvents the emission shifts to about 330 nm. This shift provides a sensitive method for detection of the transfer of tryptophan residues from the non-polar interior of a protein into contact with the solvent. So, for example, if the fluorescence of a protein solution is measured at 330 nm as a function of temperature, an unfolding curve such as that shown in Figure 5.12 will be obtained. The fluorescence decreases slightly with increasing temperature, and then increases rapidly as the protein undergoes denaturation. In this (hypothetical) case the temperature for 50% denaturation, the melting temperature T_m, is about 42 °C. Proteins vary widely in their T_m values. Some proteins, particularly those from thermophilic micro-organisms that live in inhospitable environments such as hot springs, have T_m values as high as 100 °C. The origins of this thermostablity are not well understood.

Another commonly used technique for studying protein denaturation is **circular dichroism (CD)**. The ordered secondary structures in proteins

Enzymes from thermophilic organisms have found industrial applications because they retain their activities at elevated temperatures. For example, some biological washing powders contain thermophilic proteinases and lipases. These digest the proteins and lipids in the soiled clothing.

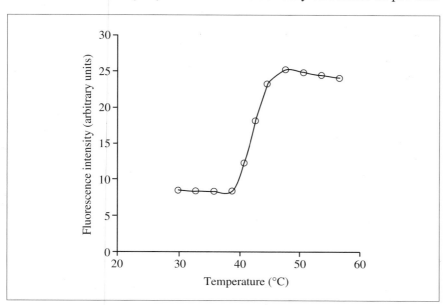

Figure 5.12 Heat denaturation curve for a hypothetical protein

are inherently asymmetric. As a result, the absorption of the left and right circularly polarized components of **plane polarized light** by peptide bonds in these ordered structures is different. Hence if plane polarized light with a wavelength around 220 nm, the region where peptide bonds absorb, is passed through a protein solution, the emergent light is **elliptically polarized**. The extent of this is referred to as the **ellipticity**, θ. A plot of θ against wavelength is a CD spectrum. The two main types of secondary structure have distinctly different CD spectra, as shown in Figure 5.13.[10] The random coil configuration has a small positive ellipticity at 220 nm. Consequently, if the ellipticity of a protein solution is measured as a function of temperature, then a curve similar to that in Figure 5.12 will be obtained.

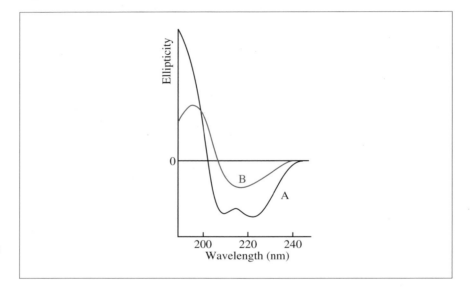

Figure 5.13 Circular dichroism spectra of the α-helix (A) and the β-sheet (B) (Adapted from Adler et al.[10])

These techniques can equally be used to study denaturation of proteins as a result of factors such as change in pH, or the addition to the solution of increasing concentrations of denaturing agents such as urea or guanidinium chloride. Generally it is found that proteins are stable over only a restricted range of pH, the precise value depending on the particular protein. Outside this range denaturation occurs, presumably as a result of changes in ionization state of acidic and basic amino acid side chains. Similarly, different proteins differ in the concentrations of urea or guanidinium chloride that cause denaturation. With urea, the effective concentrations are generally in the range 3–6 mol dm^{-3}; guanidinium chloride is effective at lower concentrations. The reasons why these agents denature proteins are not entirely clear, but it probably has to do with the structure of water in solutions of high denaturant concentration. We have already seen that the dominant factor in protein folding is the increase in entropy of water when hydrophobic side chains are sequestered into the interior of the folded protein. It is therefore reasonable to speculate that addition of high concentrations of agents that will themselves alter the properties of the solvent will have an effect on protein folding and unfolding.

Box 5.4 Free Energies of Unfolding from Denaturation Curves

Once a plot of the change of some physical property (for example, fluorescence intensity) as a function of denaturant concentration has been obtained, the equilibrium constant for the transition from native to unfolded state can be obtained. For the process shown in equation (5.1), if the fraction of the protein molecules unfolded is α, then the equilibrium constant K is given by:

$$K = \frac{\alpha}{1-\alpha} \tag{5.4}$$

Consider the example shown in Figure 5.14. The brown lines show the variation of the fluorescence intensity of the native and unfolded forms as a function of urea concentration, extrapolated from the experimental points. Suppose that we want to know the fraction of the unfolded form at the concentration indicated by the dashed brown line. It is assumed that the fluorescence intensity varies linearly with the fraction of unfolded form. Hence α is given by the ratio b/a, where b and a are as shown in Figure 5.14. The standard free energy for unfolding can then be calculated from:

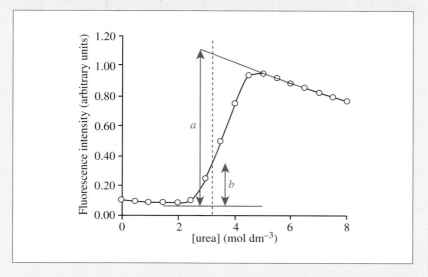

Figure 5.14 Denaturation of a protein by urea

$$\Delta G° = -RT \ln K \qquad (5.5)$$

This gives the value of $\Delta G°$ at a particular urea concentration. To obtain the value in the absence of urea it is necessary to repeat the calculation at several points on the experimental curve, and then extrapolate the values obtained back to zero concentration. Values for typical proteins, under optimum conditions of solvent, pH and temperature, fall in the range -20 to -40 kJ mol^{-1}.

5.6   Examples of How Proteins Work

As described in Chapter 1, proteins carry out a vast range of functions. In many cases, knowledge of the three-dimensional structure provides insights into how a particular protein works. In this section, a few examples have been chosen to illustrate some of the ways in which structure is related to function. The books listed under Further Reading are strongly recommended for a more extensive treatment.

5.6.1 A DNA Binding Protein

Control of the expression of the genetic information contained in DNA is essential for the proper functioning of an organism, and this control is carried out by proteins that bind to specific base sequences in the DNA molecule. An extensively studied example is a protein called Cro from the **bacteriophage** 434. This protein binds to a particular region of the phage DNA and stops expression of the gene for a protein that prevents the phage from entering what is known as the lytic phase of its life cycle. The phage then replicates and destroys the host.

Bacteriophages are viruses that infect bacteria. They are also referred to simply as phages.

The three-dimensional structure of the Cro protein bound to the piece of DNA for which it is specific is shown in Figure 5.15 (3CRO).[11] The protein is a dimer of identical subunits and is symmetric in a mirror plane perpendicular to the centre of the model shown in Figure 5.15. The interactions made by the two monomers with the DNA molecule are the same. The main interactions are with the major groove of the DNA double helix, and the shape of the Cro protein is tailored to fit into the grove. This is shown in the spacefill model at the right of the figure, where all atoms within 10 Å of the point indicated by the arrowhead have been selected. The tightness of the packing is clear. There are many interactions between the protein and the DNA in this region, but the most important for specificity of binding are hydrogen bonds made between a pair of glutamine residues and specific bases in the DNA. One of these glutamines can be clearly seen in the model. There are also interactions between Cro and the minor groove of the DNA. A spacefill model of this region is at the bottom of the figure. Again, atoms have been selected within 10 Å of the point indicated. The arrowhead points to a pair of arginines, one from each subunit, which interact with phosphates from the DNA backbone.

The essential point about Cro protein is, of course, that it binds to a piece of DNA with a specific base sequence and to no other. Nevertheless, it shares a structural feature with many other DNA-binding proteins.

Figure 5.15 Cro protein and its interaction with DNA. The spacefill models are from the regions indicated by the arrows. The cartoon shows the helix–turn–helix motif in brown

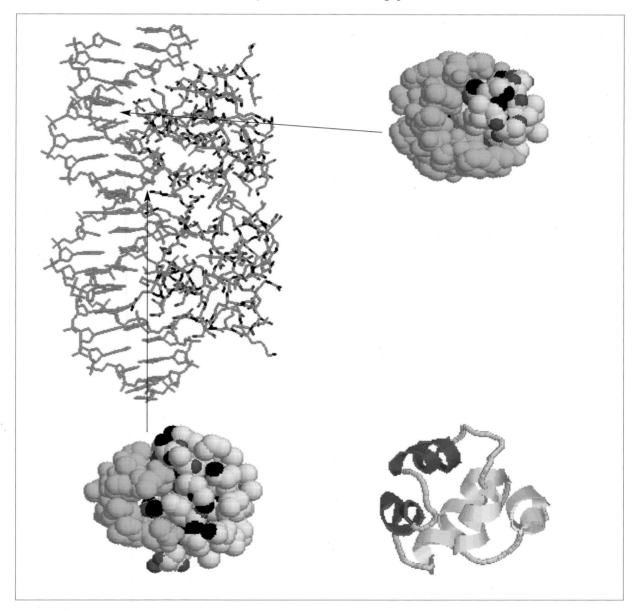

This feature is a pair of α-helices separated by a turn: the so-called **helix–turn–helix motif**. The two helices concerned are shown in brown in the cartoon at the bottom of Figure 5.15. The helix at the top is the one which binds to the major grove of DNA and the sequence of which is primarily responsible for the specificity of binding.

5.6.2 An Antibody

The outline structure of an antibody molecule (also known as an **immunoglobulin**) is shown in Figure 5.16. The molecule consists of four polypeptide chains. There are two identical **light chains** (shown in grey) and two identical **heavy chains** (shown in brown). The light chains have about 220 amino acid residues and the heavy chains have about 440. There are disulfide bridges linking pairs of light and heavy chains, and also between the heavy chains; these are indicated by grey lines. The light chain is divided into two roughly equal size **domains**, as indicated by the black horizontal lines. The heavy chain is divided into four domains; the second and third of these are separated by a short stretch of amino acids called the **hinge region**, about which the two halves of the heavy chain can bend.

The description of the structure of an antibody given here refers specifically to a subclass of immunoglobulins called immunoglobulin G, or **IgG**. There are other classes called IgA, IgE, IgD and IgM, whose structures differ in some details. The different classes of immunoglobulin have different biological functions. For example, molecules of the IgE class are involved in allergic responses such as hay fever.

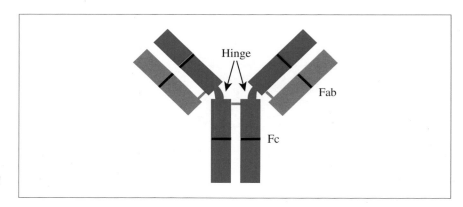

Figure 5.16 Outline structure of an immunoglobulin

The second half of the light chain has an amino acid sequence that is invariant from one immunoglobulin to another and is referred to as the **constant region**, or C_L. The sequence of the N-terminal half (at the top of the cartoon), on the other hand, is variable. It is referred to as the **variable region**, or V_L. In particular, three stretches of sequence (around residues 30, 50 and 90) vary very considerably between different molecules, and are called the **hypervariable regions**. Similarly, the three C-terminal domains of the heavy chain are constant in sequence, and are called C_{H1}, C_{H2} and C_{H3}. Just as with the light chain, the N-terminal domain is variable (V_H) and also contains three hypervariable sequences. It is these six hypervariable sequences, three from the light chain and three from the heavy chain, that come together to form the specific binding site for the antigen. Because the immunoglobulin consists of two identical pairs of chains, it has two identical binding sites. The constant regions of the molecule, particularly those of the heavy chains, are involved in biological functions other than antigen binding, but these need not concern us here.

The **antigen** is the molecule in response to which a particular antibody molecule is produced, and to which the antibody binds.

Each of the constant domains in the antibody molecule has a similar β-barrel fold consisting of a three-stranded β-sheet packed against a four-stranded β-sheet. A cartoon of this structure is shown in Figure 5.17. The strands are labelled A–G starting at the N-terminal end of the domain. Strands A, B, E and D, all running antiparallel, form one sheet. Strands G, F and C, again all running antiparallel, form the other sheet. The variable domains are basically similar except that there is an extra pair of β-strands, C′ and C″, between strands C and D. The hypervariable regions of these domains link strands B and C, strands C′ and C″, and strands F and G.

Figure 5.17 β-Barrel core of the constant domains of an immunoglobulin

If an immunoglobulin molecule is digested with the proteinase **papain**, cleavage occurs at the hinge region of the heavy chains (Figure 5.16). This yields three fragments. One consists of the third and fourth domains of the two heavy chains, linked by a disulfide bridge, and is called the **Fc fragment** (from fragment, crystallizable). The other two fragments are identical and consist of a light chain linked to the first two domains of a heavy chain; they are called **Fab fragments** (from fragment, antigen binding). The Fab fragments contain intact binding sites, and most studies of the structures of antibodies, particularly ones aimed at understanding antigen–antibody binding, have been done with Fab fragments.

An example of an Fab fragment is shown in Figure 5.18 (2MCP).[12] The light chain is on the left, and the first two domains of the heavy chain on the right. The hypervariable regions of the two chains come together at the top of the fragment to form the antigen binding site. It is the precise shapes and amino acid compositions of the six hypervariable regions within this site that provide the specificity of the antibody.

Figure 5.18 Cartoon of the Fab fragment of an antibody molecule.[12] The antigen, phosphorylcholine, is shown in spacefill

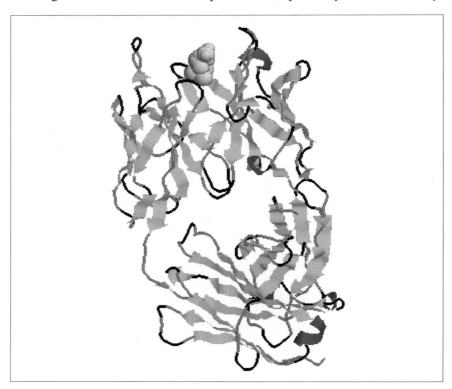

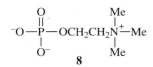

8

Antigens are usually large molecules such as proteins or polysaccharides. It is possible, however, to produce antibodies to small molecules such as dinitrophenol or phosphorylcholine. The small molecule is attached to a carrier protein and the adduct injected into the experimental animal. Antibodies are formed that recognize the small molecule. Used in this way, the small molecule is referred to as a **hapten**

In the example in Figure 5.18, the antigen is a small molecule, phosphorylcholine (**8**). It is shown in its binding site as a spacefill model. Because the binding site in the figure is shown as a cartoon, it is not possible to see why the antibody specifically binds phosphorylcholine. To make this clear, Figure 5.19 shows three views of the antigen (drawn in spacefill), and those amino acids which are within a distance of 8 Å of the antigen (drawn in wireframe). In the first part of the figure the three amino acids Tyr-33H, Tyr-100L and Trp-107H have been emphasized (H and L stand for heavy and light chain residues). These three residues form a hydrophobic pocket into which the three methyl groups of the antigen fit. In the second part, residues Arg-52H and Tyr-33H are emphasized. Both of these form hydrogen-bonded interactions with the phosphate group. Finally, in the third part, Asp-97L and Glu-35H are shown (note that only the $-CH_2CO_2^-$ part of the side chain of Glu-35H is within the specified distance from the antigen). These interact electrostatically with the positive charge on the nitrogen of the antigen. From these models it should be clear that the binding site is tailor made to fit both the shape and the charge distribution of the antigen, and moreover that both the heavy and the light chain hypervariable regions contribute to the site.

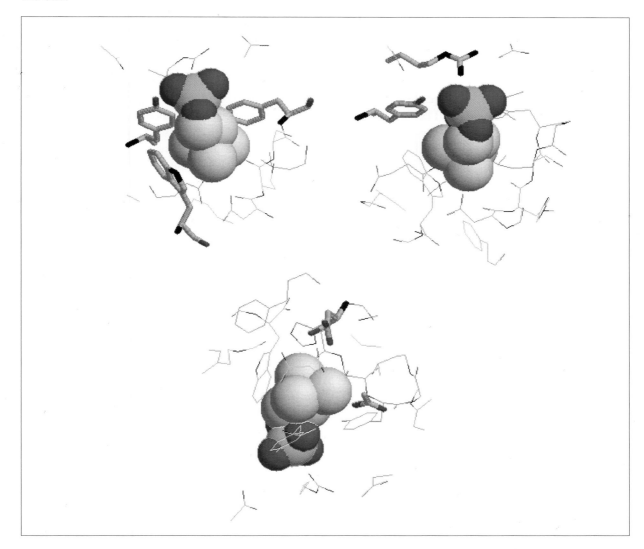

In the more usual case of protein antigens, the binding sites are much larger and make more interactions with the antigen. This provides even more scope for specificity in the binding.

Figure 5.19 Three views of the binding site for phosphorylcholine

5.6.3 An Enzyme

In the examples given above the essential feature is specific binding between the protein and some other species. In the case of enzymes, specific binding is also involved. Enzymes work on only one substrate, or a small number of related substrates, and the first step is always binding of the substrate to the active site of the enzyme. There is, however, a second essential factor. Once the substrate has bound to the enzyme, a sequence of reactions occurs that leads to rapid conversion of the substrate into products. That is, there is a catalytic process. To understand how the binding and catalysis occur again requires knowledge of the three-dimensional structure of the enzyme.

The first enzyme whose structure was determined in three dimensions was **lysozyme**. This was done by David Phillips and his co-workers.[13] The enzyme from hen egg white has 129 amino acid residues with four disulfide bridges. The natural substrate of lysozyme is the cell wall polysaccharide of certain bacteria. This polysaccharide consists of repeating units of N-acetylglucosamine (NAG) and N-acetylmuramic acid (NAM). The sugar units are β1→4 linked; a short section of the polymer is shown in **9**. Cleavage occurs at the glycosidic bond between C_1 of NAM and C_4 of NAG, at the point indicated by the arrow in **9**. Poly-NAG is also a substrate for the enzyme.

It is not, of course, possible to do X-ray crystallography on an enzyme with its bound substrate because, in the time taken to carry out the observations, the enzyme would convert the substrate to products. Indeed, the fact that enzymes are catalytically active in the crystalline state is further evidence that the structure in the crystal is the same as that in solution. Structures can, however, be obtained of enzymes with bound inhibitors or substrate analogues at the active sites. From the information obtained it is then possible to work out how the true substrate binds.

A model of the structure of lysozyme is shown in Figure 5.20 (3LYZ).[14] A prominent feature of the structure is that the enzyme has two distinct lobes separated by a cleft. The substrate binds in this cleft, and model building showed that the cleft can accommodate six monosaccharide units. Cleavage of the substrate occurs between the fourth and fifth residues of the substrate,[13] and the bond cleaved has a glutamic acid on one side and an aspartic acid on the other. Figure 5.21 shows a model of the enzyme with tetra-NAG bound in the active site (1LZC);[15] the remaining two monomers of the substrate would bind at the top of the molecule. The catalytically active glutamic and aspartic acid residues are shown in spacefill.

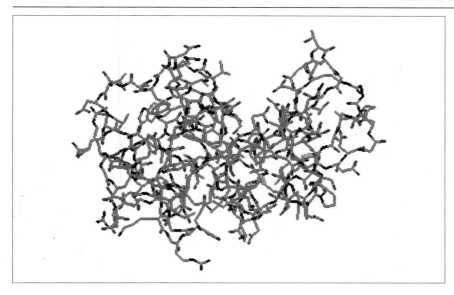

Figure 5.20 Wireframe model of lysozyme showing the active site cleft

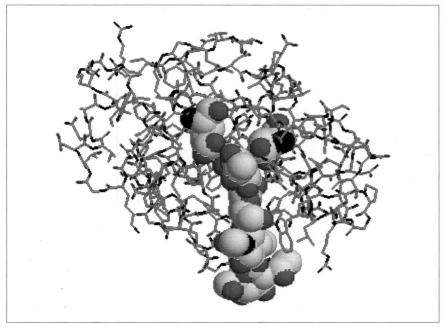

Figure 5.21 Lysozyme with tetra-*N*-acetylglucosamine in the active site cleft. Residues Glu-35 and Asp-52 are shown in space-fill

Based on the structure of the enzyme–substrate complex, and on his knowledge of the mechanism of hydrolysis of glycosides, Charles Vernon proposed a mechanism for the action of lysozyme.[16] The mechanism is outlined in Scheme 5.2. In this scheme, only two of the monosaccharide units, D and E, are explicitly shown, and the others are represented by letters; for clarity, ring substituents are not shown, but it should be borne in mind that it is the interactions between these substituents and amino acids in the cleft that are responsible for the proper location of the substrate. From the enzyme, only Glu-52 and Asp-35 are shown, one on either side of the glycosidic link between rings D and E (the cleavage point). Glu-35 is in a non-polar environment in the enzyme, and hence will be un-ionized at the pH (around 5) at which the enzyme is maximally active. When the substrate binds, Glu-35 donates a proton to its glycosidic oxygen as shown, thus making it a good leaving group. The next step is unimolecular fission of the C_1–O bond. Fragment EF then diffuses away from the enzyme and is replaced by a water molecule. Reprotonation of Glu-35 and attack of the hydroxide ion onto the carbonium ion at C_1 generates the product ABCD, which diffuses away from the enzyme.

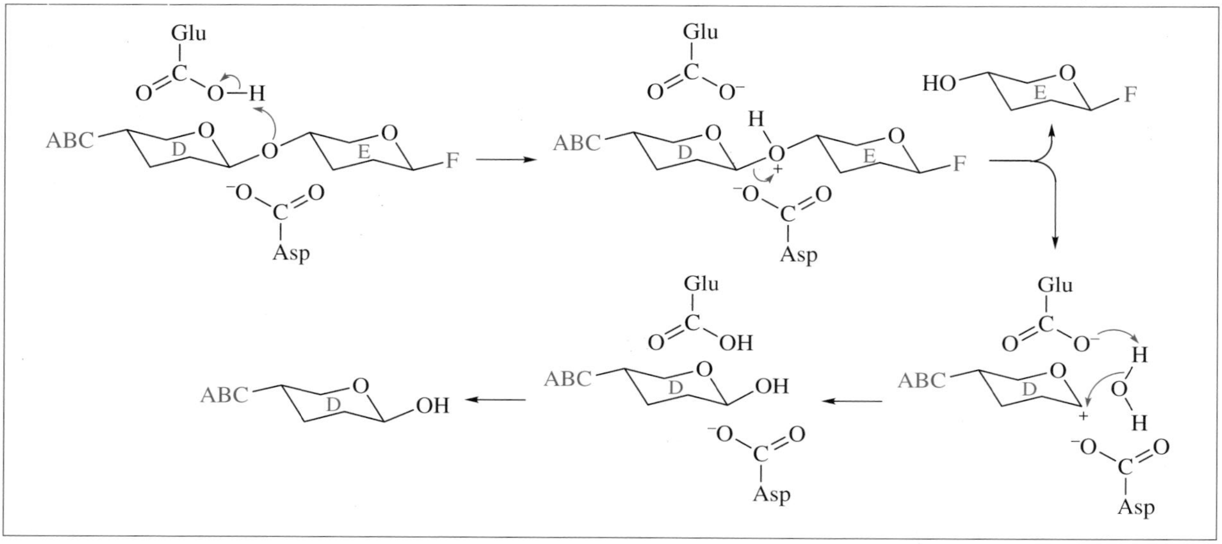

Scheme 5.2

Once the mechanism has been determined, the question arises as to why the reaction sequence is fast; that is, what factors are responsible for **catalysis**? There are three possible factors. Firstly, Glu-35 acts as a general acid catalyst and protonates the glycosidic oxygen, thereby converting it into a good leaving group. Secondly, the negative charge on Asp-52 promotes heterolysis of the C_1–O bond by stabilization of the transition state, leading to carbonium ion formation. The negative charge on Asp-52 and the carbonium ion do not collapse to form a covalent bond because they are held rigidly about 3 Å apart; this is too far for a bond to form, but close enough for a stabilizing electrostatic interaction. The third factor is not apparent from the reaction sequence as given in Scheme 5.2. Carbonium ions of monosaccharides are thought to adopt the so-called **half-chair conformation** in which C_1 is co-planar with the ring oxygen and with C_2 and C_5. This allows the positive charge to be partially located on the ring oxygen. If this is so, then the transition state leading to carbonium ion formation will have some half-chair character. Model building[13] showed that when the substrate binds to the active site, ring D is distorted towards a half-chair conformation; this will have the effect of decreasing the activation energy between the ground state and the transition state of the reaction.

Which, if any, of these factors is the most important? Lysozyme catalyses the hydrolysis of its substrate by a factor of about 10^{10}-fold compared with the same reaction in aqueous solution. This implies a decrease in the energy of activation of about 60 kJ mol^{-1}. Although the calculations cannot be done with any precision, it is likely that the energy of interaction between Asp-52 and the developing positive charge in the transition state could provide the bulk of this, with the other two factors making smaller contributions. It is not, however, profitable to push arguments of this sort too far.

Space does not allow detailed treatment of other enzyme systems here. The strategy used by an enzyme to catalyse its reaction depends on the reaction type. For example, many enzymes work by an initial nucleophilic attack on the substrate by an active site residue to form a covalent enzyme–substrate intermediate that breaks down to products. A case in point is the so-called **serine proteinases**, in which a serine residue in the active site acts as the nucleophile. The outline reaction mechanism is shown in Scheme 5.3. Here, only the essential serine residue of the enzyme is shown. In the first part of the reaction, the serine residue attacks the peptide bond to be cleaved and forms an acyl–enzyme intermediate. The C-terminal part of the substrate is released. Then a water molecule hydrolyses the intermediate to release the N-terminal part of the substrate. There are two main reasons why this reaction proceeds rapidly. First, adjacent to the serine there is a histidine residue which acts as a base to remove the proton from the serine –OH group. Secondly, bulk solvent is excluded from the active site when the substrate binds, and so the negatively charged serine is unsolvated and hence a powerful nucleophile.

From studies such as these it is possible to make some generalizations about how enzymes work. First, the active site is designed for selective substrate binding. Sometimes binding is associated with distortion of the

Scheme 5.3

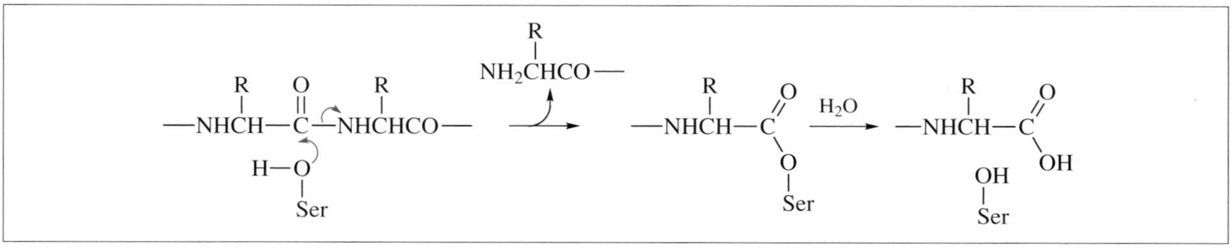

substrate towards a conformation similar to that of the transition state for the reaction. Within the active site, strategically placed amino acid side chains promote the reaction concerned. Exclusion of bulk solvent (water) from the active site creates a special microenvironment in which the reactive amino acid side chains show increased reactivity. The rest of the enzyme molecule is responsible for creation of the active site, sometimes by bringing together amino acids from remote regions of the polypeptide chain.

Summary of Key Points

- Globular proteins fold up to produce roughly spherical molecules with precisely defined shapes.
- The conformation of a polypeptide can be described in terms of two dihedral angles for each residue.
- Certain pairs of dihedral angles produce regular secondary structures, namely the α-helix and the β-strand.
- Secondary structures are stabilized by the formation of hydrogen bonds.
- The three-dimensional structure of a protein is determined by a set of weak interactions – hydrogen bonding, charge and dipole interactions, van der Waals interactions – that each residue makes with its neighbours.
- Protein folding is spontaneous, and is driven by the increase in the entropy of the solvent (water) that occurs when hydrophobic side chains are removed from the solvent into the interior of the protein.
- Proteins can be unfolded by changes in conditions such as temperature, pH and the addition of denaturing agents such as urea. The native structures of proteins are only marginally stable compared with the random coil form.
- The biological activity of many proteins depends on specific recognition of other structures by tailor-made binding sites.
- Enzymes are required to not only recognize their substrates, but also to catalyse their interconversion into products. The ways in which this is done are, in some cases, understood.

Problems

5.1. Refer to Worked Problem 5.3. Carry out the same exercise for a β-sheet with three parallel peptide chains.

5.2. In an α-helix, the rise up the helix for each amino acid residue is 1.5 Å. What is the rise for one complete turn of the helix?

5.3. State why it would not be possible to make a right-handed α-helix from D-amino acids. What helical structure might you expect to be formed in this case?

5.4. Calculation of the configurational entropy of a protein from equation (5.3) obviously requires an assumption to be made about the number of possible conformations per residue in the random coil. Test the degree of dependence of the result obtained on this assumed value by repeating the calculation in Worked Problem 5.4 with $n = 8$. Comment on the result.

5.5. Refer to Figure 5.14. From the graph, estimate the degree of unfolding of the protein at a urea concentration of 4.0 mol dm^{-3}. Assuming that the experiment was carried out at 25 °C, calculate the standard free energy of unfolding at that concentration of urea.

(Note: other problems relating to the three-dimensional structures of proteins require the use of computer techniques for their solution. Some problems of this type will be found in Chapter 6).

References

1. W. A. Hendrickson and M. M. Teeter, *Nature*, 1981, **290**, 107.
2. G. Rhodes, *Crystallography Made Crystal Clear*, Academic Press, San Diego, 1993.
3. J. N. S. Evans, *Biomolecular NMR Spectroscopy*, Oxford University Press, Oxford, 1995.
4. G. N. Ramachandran and V. Sasisekharan, *Adv. Protein Chem.*, 1968, **23**, 283.
5. S. A. Peterson, T. Klabunde, H. A. Lashuel, H. Purkey, J. C. Sacchettini and J. W. Kelly, *Proc. Natl. Acad. Sci. USA*, 1998, **95**, 12956.
6. T. Alber, D. W. Banner, A. C. Bloomer, G. A. Petsko, D. Phillips, P. S. Rivers and I. A. Wilson, *Philos. Trans. R. Soc. London, Ser. B*, 1981, **293**, 159.

7. R. Z. Kramer, J. Bella, P. Mayville, B. Brodsky and H. M. Berman, *Nat. Struct. Biol.*, 1999, **6**, 454.
8. C. B. Anfinsen, *Science*, 1973, **181**, 223.
9. D. Sheehan, *Physical Biochemistry*, Wiley, Chichester, 2000, p. 61.
10. A. J. Adler, N. J. Greenfield and G. D. Fasman, *Methods Enzymol.*, 1973, **27**, 675.
11. K. Aggarwal, D. W. Rodgers, M. Drottar, M. Ptashne and S. C. Harrison, *Science*, 1988, **242**, 899.
12. D. M. Segal, E. A. Padlan, G. H. Cohen, S. Rudikoff, M. Potter and D. R. Davies, *Proc. Natl. Acad. Sci. USA*, 1974, **71**, 4298.
13. C. C. F. Blake, L. N. Johnson, G. A. Mair, A. C. T. North, D. C. Phillips and V. R. Sharma, *Proc. R. Soc. London, Ser. B*, 1967, **167**, 378.
14. C. C. F. Blake, D. F. Koenig, G. A. Mair, A. C. T. North, D. C. Phillips and V. R. Sarma, *Nature*, 1965, **206**, 757.
15. K. Maenaka, M. Matsushima, H. Song, F. Sunada, K. Watanabe and I. Kumagai, *J. Mol. Biol.*, 1995, **247**, 281.
16. C. A. Vernon, *Proc. R. Soc. London, Ser. B*, 1967, **167**, 389.

Further Reading

C. Brandon and J Tooze, *Introduction to Protein Structure*, 2nd edn., Garland, New York, 1999.
A. Fersht, *Structure and Mechanism in Protein Science*, Freeman, New York, 1999.

6

Protein Chemistry *In Silico*

Aims

By the end of this chapter you should understand:

- That a great deal of protein chemistry is carried out using computers
- What protein databases are, what they contain, and some of the uses to which they are put
- How to use the molecular graphics program RasMol
- How to carry out protein sequence searching and comparison, and what the results mean
- How protein structures have evolved
- How protein three-dimensional structures can be predicted by homology modelling
- What the term proteomics means and how studies in this field are carried out

6.1   Introduction

Protein chemistry has traditionally been carried out either *in vivo* (in the living organism, be it animal, plant, or micro-organism), or *in vitro* (literally "in glass", or in the test tube). Much of modern protein chemistry is carried out *in silico*, that is, within the microchip of a computer. The reason for this is that there is now such a wealth of sequence and structural data available that it is impossible to store and analyse it without the aid of computers. So, for many protein chemists, the computer is the laboratory where the experiments are done, and **databases** are the equivalent of the stores from which materials are withdrawn to do the experiment. This chapter is concerned with what some of those databases contain, how the information is retrieved, and what sort of things can be done with it. This is the field of what has become known as **bioinformatics**.

To get the most out of what follows, and to do some of the problems, you will need access to a computer connected to the internet. Modern PCs are sufficiently powerful to do a considerable amount of work in bioinformatics, and all of the applications described here were run on such a machine. For more advanced applications, particularly when a large amount of computer power is needed, access to a dedicated workstation is required. This will almost certainly be running under the Unix operating system. It is possible to mount Unix (or the open source software version Linux) on a PC, and even to partition the hard drive so that both Windows and Unix can run on the same machine. If you want to do that, then a recent book by Gibas and Jambeck[1] is an invaluable guide.

Freeware is the term used for computer software, the executable version of which is available free of charge. **Open source software** describes programs for which the **source code** is also available, so that expert programmers can modify it to suit their own needs.

6.2 Protein Databases

Protein databases are much more that just a repository of structural information. For a start, the experimental results that are submitted to databases are, as far as possible, checked for correctness before being deposited in the database. They are also annotated to give background information about the protein in question, and hyperlinks are provided to other important data sources. Moreover, databases also contain programs that can be used to extract the data that one requires, and to

Table 6.1 Some databases important for protein chemistry

Site	Contents	Address
European Bioinformatics Institute (EBI)	Protein sequences, 3-D structures, analysis tools	http://www.ebi.ac.uk/
Swiss Institute of Bioinformatics (SIB)	Protein sequences, 3-D structures, analysis tools	http://www.expasy.ch/
Research Collaboratory for Structural Bioinformatics (RCSB)	3-D structures (Protein Data Bank)	http://www.rcsb.org/index.html
National Centre for Biotechnology Information (NCBI)	Entrez search and retrieval system	http://www.ncbi.nlm.nih.gov/
Protein Information Resource (PIR)	Protein sequences, analysis tools	http://pir.georgetown.edu/

analyse it. Indeed, it is often unnecessary to download the structures of the molecules in which one is interested – the analyses can be done online using programs within the database.

For the present purposes, the most important databases are those listed in Table 6.1. They all contain a vast amount of information, and the only way to get a full appreciation of their contents is to go in and look around. **ExPASy** at the Swiss Institute of Bioinformatics probably offers the most comprehensive set of tools for analysis of proteins. The main database for three-dimensional structures is that maintained by the Research Collaboratory for Structural Bioinformatics at Rutgers University. This started life in 1971 at the Brookhaven National Laboratory, and is sometimes (incorrectly) still referred to as the **Brookhaven database**. The National Centre for Biotechnology Information in Bethesda, Maryland, is now more important as a database for DNA sequences and genomes than for proteins. It does, however, have a search and retrieval system called **Entrez** which is extremely useful for sourcing information. The **Protein Identification Resource** evolved from a sequence database started by the late Margaret Dayhoff, who was one of the first people to recognize that storage and analysis of the rapidly accumulating protein structural data would only be feasible by electronic means.

In the examples of database use that follow, reference will be made to particular sites, but problems can usually be solved using the facilities at more than one site, and to some extent the one used is a matter of personal preference.

There are many other specialized databases that contain information of interest to the protein chemist, mainly information derived by analysing primary structural data. Examples are **SCOP** and **CATH**, which group proteins into three-dimensional structural classes. These can be located using any of the standard search engines.

Worked Problem 6.1

Q Retrieve the amino acid sequence of the peptide melittin from the honeybee.

A There is no single way to do a database search. The following is one possibility. Log on to the NCBI. Click on *Entrez*, select *Protein* as the field to search, and enter *melittin* as the search term. This will produce (at the time of writing) 68 entries; which do we want? One approach is to look through them all to find one specifically related to melittin from the honeybee. It is better to refine the search using *melittin honeybee* as the search term; this searches for a record containing both terms and produces one entry with the identification code S23131. Clicking on that produces the following record:

```
LOCUS   S23131      26 aa     linear INV 19-MAR-1997
DEFINITION  melittin – honeybee.
ACCESSION  S23131
PID   g283738
VERSION  S23131  GI:283738
DBSOURCE  pir: locus S23131;
   summary: #length 26  #molecular-weight 2847  #checksum
   7439; PIR dates: 19-Mar-1997  #sequence_revision 19-Mar-
   1997  #text_change
   19-Mar-1997.
KEYWORDS.
SOURCE  honeybee.
   ORGANISM  Apis mellifera
   Eukaryota; Metazoa; Arthropoda; Tracheata; Hexapoda;
   Insecta; Pterygota; Neoptera; Endopterygota; Hymenoptera;
   Apocrita; Aculeata; Apoidea; Apidae; Apis.
REFERENCE  1 (residues 1 to 26)
   AUTHORS  Ramalingam, K. and Bello, J.
   TITLE  Effect of permethylation on the haemolytic activity of
      melittin
   JOURNAL  Biochem. J. 284 (Pt 3), 663–665 (1992)
   MEDLINE  92321983
FEATURES    Location/Qualifiers
   source  1..26
         /organism="Apis mellifera"
         /db_xref="taxon:7460"
   Protein  1..26
         /product="melittin"
ORIGIN
    1 gigavlkvlt tglpaliswi krkrqq
Bottom of Form
Revised: October 24, 2001.
```

The amino acid sequence is in the penultimate line in lower case,
single-letter code. There is obviously a lot more information in the
record. At the top is confirmation of the identity of the peptide,
followed by details of which database it was obtained from (PIR).
It also gives the length (26 aa), and the M_r (2847). Then there is
information about the taxonomy of the honeybee, *Apis mellifera.*
Note that the term *Apis mellifera* is highlighted (blue in the original record). This is a **hyperlink** which, if clicked, will take you to
further information about the honeybee. Next there is information

about the authors of the work and the journal where it was pub-
lished. MEDLINE is a literature database, and clicking the hyperlink shown will take you to an abstract of the publication cited.
Finally, there is the sequence and a note of the last occasion on
which the record was revised.

6.3 Molecular Modelling

Pictures of models of protein structures, such as those in Chapter 5, are
limited in their ability to provide information about the important structural features of a protein. They provide a single, flat view. It is possible to construct stereo views which, with practice, allow the model to be
seen in three dimensions (see Worked Problem 6.3). However, to appre-
ciate fully what a protein structure looks like, and even more to obtain
answers to questions about its features, there is no practical alternative
to using a molecular modelling program to draw structures on a computer screen. The structure can then be rotated to see it from any direction, parts of it can be selected for more detailed views, distances between
features can be measured, and so on.

The molecular graphics program used to produce the images in this
book is **RasMol**. This program was created by Roger Sayle, who made
it freely available to the scientific community. It is now maintained at
the University of Massachusetts. You are strongly urged to obtain a copy
and install it on your computer; otherwise you will not be able to do
some of the examples that follow.

First create a directory to put the program in (*e.g.* C:\RasMol), and
then go to the RasMol home page at http://www.umass.edu/microbio/
rasmol/, where you will find a description of the program and a link to
take you to the download page. You will need to look through the information on that page to decide which version of the program you need
(which depends on the characteristics of your computer system), but it
is then just a matter of downloading it into the directory you have created, and then installing it. Be sure to download the help files and the
RasMol reference manual at the same time.

Box 6.1 Protein Explorer

The RasMol home page also describes another program called
Protein Explorer. This is a development from RasMol and, in some
ways, is more powerful. For example, it allows two different mol-

ecules to be displayed at the same time, which is useful for comparative purposes. It is designed to run inside a Web page and is extremely useful when visiting a site such as RCSB, because it can be used to view molecules without downloading them. To use it you will need a web browser plug-in called Chime. This is freely available from its creators (http://www.mdlchime.com). There is a snag, however. At the time of writing, Chime will not run in Netscape 6, so if you use Netscape 6 you cannot run Protein Explorer. You need Netscape 4.7x, which can be downloaded from the Netscape website. Protein Explorer is great fun to use, so it is worth the effort of getting it working on your machine.

You will obviously need some structures to draw with RasMol. These can be downloaded from the database of the RCSB (or from other sites listed in Table 6.1).

Worked Problem 6.2

Q Retrieve the atomic coordinates of crambin from the RCSB web site.

A This is straightforward because we already know the identification code (1CRN; see p. 109) for the protein the coordinates of which we wish to retrieve. Go into the web site and then the PDB (Protein Data Base) page. In the *Find Structure* box enter *1CRN* and *go*. A list of entries will appear including 1CRN. Click on *EXPLORE*. You will get a screen giving summary information about the structure, and with a set of clickable options. Click on *Download/Display File*. In the *Download* box, click the *X* under *PDB/None*, and choose *Save to Disk*. Specify the address of your RasMol directory and the file will be downloaded. Once you have downloaded the file, you can examine its contents by opening the file in a word processing package (it is too large to reproduce here). It starts with a lot of background information (similar to that for a sequence file). Then there is information about which residues are involved in disulfide bridges, and which residues form part of secondary structures. Finally, there is a list of the type and the x,y,z coordinates of every atom in the molecule except the hydrogens.

(Note: if the problem had been to find the coordinates for a molecule for which the identification code was not already known, then we would enter an appropriate key word in the initial *Find Structure* box. Be prepared for a lot of entries. For example, using *Insulin* as a search term produces more than 100 records. Remember also that if you have installed Chime, then you can view the molecule before downloading. From the *Summary Information* page select *View Structure*. Then click on *Protein Explorer* and follow the prompts.

The atom type and its x,y,z coordinates are the raw material that RasMol uses to draw the molecule. The computer screen is treated like a piece of graph paper with the x-axis horizontal and the y-axis vertical. The x,y coordinates of each atom are then used to determine where that atom (or the bond drawn to it) is placed on the screen. The type of the atom (carbon, oxygen, *etc.*) determines its colour. The z coordinate is effectively ignored; that is, the molecule is flattened onto the plane of the screen. It becomes important, however, if the molecule is rotated, because where the atom ends up on the x/y plane depends on how far it was above or below it before rotation.

To begin modelling, click on the RasMol icon that will have been set up when the program was installed. This opens a window with a black drawing surface. At the top of the window are several pull down menus. Left clicking on *File*, then on *Open* produces a window listing whatever files are in the directory that are readable by RasMol – there should be at least one, 1CRN. Clicking on that file results in a wireframe model of crambin. The molecule can be translated along the x- and y-axes by right clicking and moving the mouse left/right and up/down, respectively. Left clicking and moving the mouse left/right or up/down rotates the molecule around the y- and x-axes, respectively. Rotation around the z-axis is done by holding down the shift key whilst right clicking and moving the mouse left/right. Holding down the shift key whilst left clicking and moving the mouse up/down changes the size of the molecule.

Besides these basic mouse operations, much can be done with the pull-down menus, including changing the colour scheme of the molecule, and selecting between display modes (wireframe, ball and stick, spacefill, *etc.*). However, at the bottom of the screen will be found a button labelled *RasMol Command Line*. Clicking on this opens a new window into which commands can be typed at the *RasMol>* prompt. The real power of RasMol is only obtained by making full use of this facility. The possible commands and what they do can be discovered from the user manual under the *Help* pull down, or from the reference manual. There is no substitute for studying these sources of information and trying the commands out. Some examples of commands and their use will be found in the following worked problem.

Worked Problem 6.3

Q Produce a stereo view of the β-turn from crambin in the orientation shown in Figure 5.9.

A Open file 1CRN in RasMol. Open the *Command Line* window and type in the following commands at the *RasMol>* prompt, pressing *Enter* after each.

background white (changes the colour of the drawing canvas)

wireframe 50 (changes the thickness of the bonds; the units are internal RasMol units)

select *.n?? (selects all the nitrogen atoms in the molecule; any type of atom can be selected this way. Once a group of atoms or residues has been selected, subsequent commands affect only that group. Another group or all of the residues (*select all*) can be selected at any time).

color black (changes the colour of the nitrogen atoms. Note spelling of *color* in the command. Some colours, like black, are pre-defined. Any others can be chosen by entering *color [x,y,z]* where *x*, *y* and *z* are whole numbers between 0 and 250)

restrict 41-44 (displays only residues 41 to 44. The rest of the molecule is still present, but cannot be seen)

hbond (draws any hydrogen bonds between the residues displayed. Also reports the total number of hydrogen bonds in the molecule – 27 in this case)

At this stage the display will show the β-turn with its single hydrogen bond, but not oriented as in Figure 5.9. Use the mouse to re-orient it as nearly as possible to the view in Figure 5.9. After that, go to the *Options* pull down and click on *Stereo*. This will produce two copies of the structure, one oriented a few degrees around the *y*-axis with respect to the other. To obtain a hard copy of the image, go to the *Export* pull down and select *GIF* as the file type. Choose a suitable destination for the file. The exported GIF file can be inserted into a word processor document. The figure will appear as in Figure 6.1.

The point of stereo views is that, by a technique known as *naked eye stereopsis*, they allow the structure to be seen in three dimensions. To do this, look at the page between the two models and let your eyes go out of focus. You will see copies of the two images float towards one another. When they superimpose, the combined

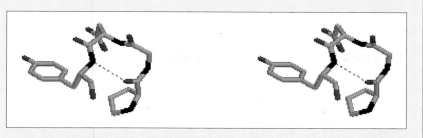

Figure 6.1 Stereo image of the β-turn from crambin

image will suddenly spring out into three dimensions. It takes a little practice to get the effect first time, but after that it is easy. However, for those who cannot do this, stereo viewers are commercially available.

6.4   Sequence Searching and Comparison

Modern methods of DNA sequence analysis are extremely rapid, and are providing an embarrassment of riches. Sometimes the sequences of genes coding for proteins of known function are determined, but often sequences are obtained that code for proteins of unknown function. This is particularly true where sequences of complete genomes are determined; the functions of many of the proteins that are encoded by the genome will be unknown. The solution to this problem comes from the fact that proteins with similar functions have related amino acid sequences (see Section 6.5). Hence the approach is to search a database for proteins whose sequences are related to that of the newly determined sequence. The main program used for this purpose is called **BLAST**[2] (from **B**asic **L**ocal **A**lignment **S**earch **T**ool). This program takes a query sequence, uses it to search the contents of a specified database, and produces a report of those proteins in the database that have structures related to that of the query sequence. An example of the use of BLAST to identify the function of an unknown protein is given in the following worked example.

Worked Problem 6.4

Q A piece of DNA from the mushroom *Armillaria mellea* was sequenced and translated into the corresponding protein. The protein sequence obtained (164 residues) was as follows:

AGPDFDLDYRTYPQSSENICYSWFCNNGPHSVAP
DRTHAAAHRASNSCGNVNPNRCSIRVGHVSGYQCDEWP
WANSNAGGANAATRCIPTADNTGSGSQWGNFINNRGSQ

AVGYVLQDNVVFATIEISNIPTTAEFCKGVLGTAITATM
CRQVANGQPYLQRIG

Carry out a BLAST search of the SwissProt database in an attempt to assign a function to this protein.

A Log in to the SIB website. On the home page, go to *Proteomics Tools* and select *Similarity Searches*. From the new page that opens up select *BLAST*. On the BLAST page, select *blastp* as the program to use (this searches a protein sequence against a protein database; other varieties of BLAST search nucleic acid sequences against nucleic acid sequence data bases, translated nucleic acids against protein databases, and so on). Select *SwissProt* as the database. Leave the next three fields at the default values shown. Type or paste the sequence given into the sequence box. It is best to type the sequence in advance using a word processor, copy it, and paste into the box; this saves time. (Note that if you are using a sequence that is already deposited in a database as the query, then all that is necessary is to enter the database ID or accession number in the box – the program will retrieve the sequence for you). Enter your e-mail address and click on *RUN*. This will open a new page confirming the details of the request that you have made and saying that the results will be e-mailed to you. An edited version of the output is shown in Figure 6.2.

The first point to note is that the SwissProt database, at the time this search was done, contained 107,093 sequences with a total of 39,364,193 letters (residues)! The results proper start about halfway down with a list of sequences producing significant alignments. Each entry gives the identity of the matching protein, and a measure (the E value) of the significance of the match. Values less than 1 are usually of interest. Only the first three matches have been retained in Figure 6.2; seven more were reported but have been deleted because their significance was low. Next comes partial sequence alignments for the matches; only the first and third have been kept. The results show the part of the query sequence for which the best match was obtained. In the first case the match starts at residue 65 of the query and aligns with the sequence of protein P42983 from residue 83. The middle line shows residues that are identical between the two sequences, and a + is put between residues that, although not identical, are very similar in type (*e.g.* Ile and Val). The string of Xs in the query sequence indicates that the residues occurring in that part of the protein are of the most commonly occurring type (Ala, Ser, Asn, Gly) and so they have not

When making a copy of the output from any sequence comparison program using a word processor you should use **Courier New** as the font. The reason is that in this font all characters are of the same size. With fonts for which this is not the case, the sequence alignment is lost.

```
Subject: Blast2.0 search results

SIB BLAST network server version 1.4 of October 6, 2001

Welcome to the SIB BLAST Network Service (sib-blast)

=====================================================================

        Swiss Institute of Bioinformatics (SIB)
        Ludwig Institute for Cancer Research (LICR)
        Swiss Institute for Experimental Cancer Research (ISREC)

=====================================================================

BLASTP 2.2.1 [Jul-12-2001]

Reference: Altschul, Stephen F., Thomas L. Madden, Alejandro A. Schaffer,
Jinghui Zhang, Zheng Zhang, Webb Miller, and David J. Lipman (1997),
"Gapped BLAST and PSI-BLAST: a new generation of protein database search
programs",  Nucleic Acids Res. 25:3389-3402.

Query 164 letters

Database: swiss 107,093 sequences; 39,364,193 total letters

                                                           Score     E
Sequences producing significant alignments:               (bits)  Value

sp|P42983|NUCB_BACSU (NUCB)Sporulation-specific extracellular nu...  31   0.76
sp|O54898|CCAG_RAT (CACNA1G)Voltage-dependent T-type calcium cha...  30   1.3
sp|P12667|NUCA_BACSU (NUCA..)DNA-entry nuclease (EC 3.-.-.-) (Co...  30   2.2

>sp|P42983|NUCB_BACSU (NUCB)Sporulation-specific extracellular
             nuclease precursor (EC 3.-.-.-).[Bacillus subtilis]
             Length = 136

 Score = 31.2 bits (69), Expect = 0.76
 Identities = 14/38 (36%), Positives = 17/38 (43%)

Query: 65   GYQCDEWPWXXXXXXXXXXXXTRCIPTADNTGSGSQWGN 102
            GY DEWP         R + +DN G+GS  GN
Sbjct: 83   GYDRDEWPMAVCEEGGAGADVRYVTPSDNRGAGSWVGN 120


>sp|P12667|NUCA_BACSU (NUCA..)DNA-entry nuclease (EC 3.-.-.-)
             (Competence-specific nuclease).[Bacillus subtilis]
             Length = 147

 Score = 29.6 bits (65), Expect = 2.2
 Identities = 24/91 (26%), Positives = 34/91 (36%), Gaps = 11/91 (12%)

Query: 12   YPQSSENICYSWFCNNGPHSVAPDRTHAAAHRASNSCGNVNPNRCSIRVGHVSGYQCDEW 71
            YP+++++I +   N G  V     A R  S +V +      GY DEW
Sbjct: 52   YPETAKHIKDA--INEGHSEVCTIDRDGAEERREQSLKDVPSKK---------GYDRDEW 100

Query: 72   PWXXXXXXXXXXXXTRCIPTADNTGSGSQWGN 102
            P             I ADN G+GS G+
Sbjct: 101  PMAMCKEGGEGASVEYISPADNRGAGSWVGH 131
```

Figure 6.2 Edited output from a BLAST search

been taken into account in the analysis; a match between the two sequences in that region might well occur by chance.

The point of all this is that the matching protein in the database is a nuclease; that is, an enzyme that hydrolyses DNA. Hence our unknown protein is almost certainly a nuclease. Subsequent experimental evidence obtained with the purified protein showed that this was indeed the case.[3] Database searching has allowed us to assign a function to this protein.

BLAST does not attempt to match up the full sequences of the query and subject proteins. It simply looks for the most similar regions. This is to increase its speed of operation. Once one or more matching sequences are obtained, it is necessary to compare the sequences in more detail. The program CLUSTAL W[4] provides a tool for doing this. CLUSTAL W is a very powerful program because it can do multiple alignments; that is, alignment of three, or more, sequences. It can either be run online, or can be downloaded onto one's own computer. For example, in Worked Problem 6.4 it was found that a query sequence was, somewhat distantly, related to a nuclease from another organism. A second nuclease (SwissProt accession code P12667) was more distantly related (see Figure 6.2, third entry in the list of sequences producing significant alignments), but the matches in the region of residues 88–102 of the query looked very similar to those found with protein P42983. A multiple sequence alignment between these three proteins would be the next move to make, and is the subject of the following Worked Problem.

Worked Problem 6.5

Q Obtain a multiple sequence alignment for the nuclease from *A. mellea* and the two proteins identified as related to it in Figure 6.2.

A The first thing to do is to download the sequences of the two matched proteins. To do this, go into SwissProt. Use the accession codes (P42983, P12667) as the search terms. In each case an entry containing a large amount of information about the protein will be obtained (not shown here). At the end of the record is a link *FastA format*. Click on this, and the sequence will be presented in the format required for FastA and for CLUSTAL W. This has a title line starting with >, followed by lines of sequence. Highlight the entry, copy it, and paste into a Word document. Similarly, put a title line onto the sequence of the protein from *A. mellea* (*e.g.* >A. mellea).

Go into the EBI website and, in *Tools*, find and click on *CLUSTAL W*. Enter your e-mail address as requested (although the results will be shown on-line). Leave all the entries in the boxes as the default values. Paste the three sequences into the sequence box (you can get online help about the format if you are unsure). Click on *Run*. If you have made any mistakes in data entry, then you will be told what they are, and what to do about it. Otherwise, the results will appear on the screen. The results can be saved, or copied from the screen. The essential parts of the results in the present case are shown in Figure 6.3.

```
sp|P42983|NUCB_BACSU    -----MKKWMAGLFLAAAVLLCLMVP------QQIQGASSYDKVLYFPLSRYPETGSHIR  49
sp|P12667|NUCA_BACSU    MTTDIIKTILLVIVIIAAAAVGLIKGDFFSADQKTSQTKEYDETMAFPSDRYPETAKHIK  60
A.mellea               ----------------AGPDFDLDYR------TYPQSSENICYSWFCNNGPHSVAPDRTH  38
                                       *.   .  *          . :..       . :. :  .: :

sp|P42983|NUCB_BACSU    DAIAEGHPDICTIDRDGADKRREESLKGIPTKPGYDRDEWPMAVCEEGGAGADVRYVTPS  109
sp|P12667|NUCA_BACSU    DAINEGHSEVCTIDRDGAEERREQSLKDVPSKKGYDRDEWPMAMCKEGGEGASVEYISPA  120
A.mellea               AAAHRASNSCGNVNPNRCSIR-------VGHVSGYQCDEWPWANSNAGGANAATRCIPTA  91
                         *   ..  .  .::  :  ..  *        :    **:  **** *  .:  **   .* ..  :..:

sp|P42983|NUCB_BACSU    DNRGAGSWVG-------NQMSSYPDGTRVLFIVQ------------------------  136
sp|P12667|NUCA_BACSU    DNRGAGSWVG-------HRLTDYPDGTKVLFTIQ------------------------  147
A.mellea               DNTGSGSQWGNFINNRGSQAVGYVLQDNVVFATIEISNIPTTAEFCKGVLGTAITATMCR  151
                       ** **  *:**    *           :    .*       .*:*

sp|P42983|NUCB_BACSU    -------------  
sp|P12667|NUCA_BACSU    -------------  
A.mellea               QVANGQPYLQRIG  164
```

In Figure 6.3, each line starts with an identifier for the protein whose sequence follows. The sequences of the three proteins have been aligned so as to maximize the number of places (shown by *) at which the same amino acid occurs in all three. Very similar, but not identical, residues are marked with a colon, and more distantly related residues with a full stop. At various points it is necessary to put gaps in one or other of the sequences (indicated by -) to optimize the alignment. These results confirm that the three proteins are members of the same family (their sequences are more similar than would occur by chance), and confirms that the protein from *A. mellea* is indeed a nuclease. The highest degree of similarity is in the region of residues 65 to 101 of the *A. mellea* protein (residue numbers are given at the end of each line). In this stretch, there are 16 identical residues. The high degree of similarity probably implies that this region of the enzyme is important for its function, and it may represent the active site.

Figure 6.3 Multiple alignment of three proteins using CLUSTAL W

6.5   Protein Evolution

One important application of protein sequence databases and protein sequence comparisons is the study of **protein evolution**. There are two distinct aspects to this. Firstly, comparison of the sequences of the same protein from different organisms provides information about the **evolution of species**. Secondly, comparison of the sequences of different proteins within the same organism provides information about the **evolution of protein function**. These two aspects will be treated briefly here; for a detailed account, see Pathy.[5]

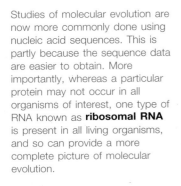

Studies of molecular evolution are now more commonly done using nucleic acid sequences. This is partly because the sequence data are easier to obtain. More importantly, whereas a particular protein may not occur in all organisms of interest, one type of RNA known as **ribosomal RNA** is present in all living organisms, and so can provide a more complete picture of molecular evolution.

If a pair-wise comparison of the same protein from two species is carried out, it is found that the sequences are (usually) different. An example is shown in Figure 6.4. The sequences of a small protein called **cytochrome** *c* from humans and from the rattlesnake (*Crotalus atrox*) were obtained from SwissProt and then aligned using CLUSTAL W. The usual symbols are used in the comparison; that is, * for identity, a colon for very similar residues, and a full stop for somewhat similar residues. The format of the report looks somewhat different from that in Figure 6.3 because this time CLUSTAL W was run at the SIB site.

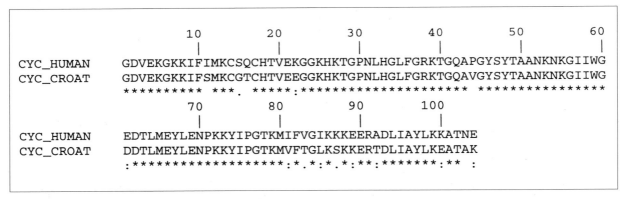

Figure 6.4 Comparison of the sequences of cytochrome *c* from humans and rattlesnake

It interesting to note that the entry for human cytochrome *c* in SwissProt reports that it is also the sequence for the cytochrome *c* from the chimpanzee (*Pan troglodytes*). As judged from the sequences of these two proteins, humans and chimpanzees have not yet diverged significantly from one another!

Of the 104 residues in the protein, 89 (85.6%) are identical between the two sequences. Cytochrome *c* sequences are available from a very large range of organisms. If pair-wise comparisons are carried out on all of them, it is found that the closer are two organisms on the evolutionary time scale, the more similar are the sequences of their cytochrome *c* molecules.

Moreover, from knowledge of the time of divergence of organisms whose ancestors are well represented in the fossil record, it is possible to determine the rate at which the sequence of cytochrome *c* has changed over time. The value obtained is about one change in sequence per 100 residues in 21.4 million years, and appears to have remained constant over time. This means that changes in sequence of a protein can be used as a **molecular clock** to time evolutionary events. It also means that **molecular phylogenies** can be constructed; that is, evolutionary trees based solely on protein sequence comparisons.[6] Molecular phylogenies are more complete than those based on the fossil record. This is because many more data are available, and because it is possible to study organisms whose ancestors have not left fossils. For these reasons, phylogenetic trees based on molecular data are now regarded as giving the most comprehensive and reliable view of the evolutionary process.

This is known as the **unit evolutionary period**. It differs from one protein to another. For example, the unit evolutionary period for globins is about 6.1 million years.

Worked Problem 6.6

Q Calculate the time that has elapsed since humans and rattlesnakes last shared a common ancestor, based on comparisons of the sequences of their cytochrome c.

A The sequences are 85.6% identical; that is, they differ by 14.4%. However, the sequences have been changing independently since divergence from the common ancestor, so each has accumulated 7.2% of change, or 7.2 changes per 100 residues. The time to accumulate 1 change per 100 residues is 21.4 million years, so the elapsed time is 154 million years. This is at the time of the border between the Jurassic and the Cretaceous periods.

In fact, it is not quite as easy as that. The possibility has to be allowed for that a given residue might have changed, and then changed back to the original one; two mutations would have occurred but none would be observed. This problem was analysed by Kimura,[7] who showed that the average number of changes per site between two sequences, K_{aa}, is related to the fraction of sequence differences, p_d, by equation (6.1):

$$K_{aa} = -\ln (1 - p_d - 0.2p_d^2) \qquad (6.1)$$

In the case under consideration, $p_d = 0.144$, so $K_{aa} = 0.160$. So 16% of changes have accumulated, 8% in each sequence, and the elapsed time is 171 million years. This more detailed analysis does not make much difference in the present case because the divergence is relatively recent. It becomes much more significant if distantly related sequences are compared.

> ## Box 6.2 The Mechanism of Protein Evolution
>
> DNA is constantly subjected to environmental insults that can result in **mutation**; that is, the change of one of its four bases into another. This may not result in a change of amino acid sequence in the coded protein because, as described in Section 4.5, the code is degenerate. Some mutations will, however, result in a change in the amino acid at a particular position. Frequently, such a change will result in a defective protein, and the mutation will be eliminated from the population. Sometimes, however, the change will be either neutral or beneficial. In such cases, there is a finite chance that the changed protein will become fixed in the population; that is, the protein will have mutated. This topic is treated in detail by Kimura.[7]

The other aspect of interest is evolution of protein function. It is reasonable to assume that the earliest living organisms contained a small number of relatively simple protein molecules. Modern day organisms contain a large number of different proteins, many of them with complex structures. How did this happen?

Comparing the sequences of different proteins from within the same organism shows that they fall into families with related sequences. In Section 5.6.3 the mechanism of action of the serine proteinases was briefly described. In fact, there are many serine proteinases, all of which have the same mechanism of action, but which differ in substrate specificity; that is, they hydrolyse proteins at different sites. Some of them will hydrolyse almost any protein wherever the site in question occurs; for example, trypsin will catalyse hydrolysis on the C-terminal side of virtually all lysine and arginine residues in any protein. Some of them hydrolyse at only a single type of site in a single protein molecule. For example, the enzyme **thrombin** is involved in the process of blood clotting. It catalyses the hydrolysis of only one substrate, **fibrinogen**, and this only at four particular Arg–Gly bonds, to produce **fibrin**. The fibrin then produces the clot.

Consider, as an example, the two digestive enzymes trypsin and chymotrypsin. The amino acid sequences of their precursor molecules, trypsinogen (SwissProt P07477) and chymotrypsinogen (SwissProt P17538), from humans are compared in Figure 6.5. As before, the alignment has been done with CLUSTAL W and the usual symbols are used. It turns out that the sequences are 38% identical. Moreover, the catalytic serine and histidine residues (shown in brown) are in identical places in the two sequences, in regions where the structures are more similar than elsewhere. There is a third important residue, an aspartic acid, whose

Many enzymes are biosynthesized as inactive **precursors** or **proenzymes** that only become converted to the active forms when they reach their site of action in the organism. This is usually to protect the cells in which they are made from undesirable activity – it is not a good idea to have digestive proteinases such as trypsin and chymotrypsin around inside a cell. In the case of trypsinogen, activation involves removing the first 23 amino acids from the N-terminus. With chymotrypsin, the first 18 residues are removed, followed by some other changes that need not concern us.

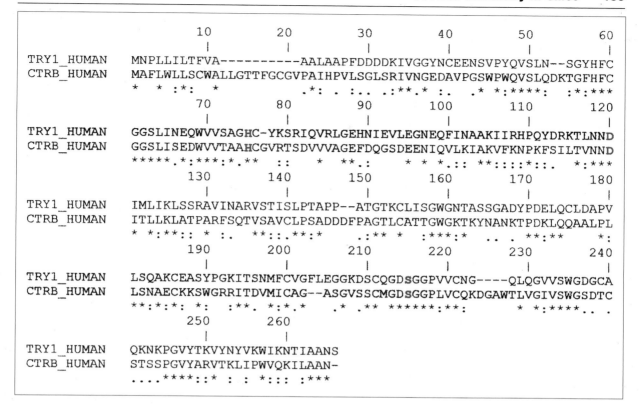

```
                    10          20          30          40          50          60
                    |           |           |           |           |           |
TRY1_HUMAN  MNPLLILTFVA----------AALAAPFDDDDKIVGGYNCEENSVPYQVSLN--SGYHFC
CTRB_HUMAN  MAFLWLLSCWALLGTTFGCGVPAIHPVLSGLSRIVNGEDAVPGSWPWQVSLQDKTGFHFC
            *    *  :*:    *           .*: . :.. .:**.* :.   .* *:****: :*:***
                    70          80          90          100         110         120
                    |           |           |           |           |           |
TRY1_HUMAN  GGSLINEQWVVSAGHC-YKSRIQVRLGEHNIEVLEGNEQFINAAKIIRHPQYDRKTLNND
CTRB_HUMAN  GGSLISEDWVVTAAHCGVRTSDVVVAGEFDQGSDEENIQVLKIAKVFKNPKFSILTVNND
            *****.*:***:*.**   ::    *   **.:     * * *.::  **::::*::.  *:***
                    130         140         150         160         170         180
                    |           |           |           |           |           |
TRY1_HUMAN  IMLIKLSSRAVINARVSTISLPTAPP--ATGTKCLISGWGNTASSGADYPDELQCLDAPV
CTRB_HUMAN  ITLLKLATPARFSQTVSAVCLPSADDDFPAGTLCATTGWGKTKYNANKTPDKLQQAALPL
            * *:**:: * :.    **::.**:*   .:** *  :***:*  .. .  **:**   *:
                    190         200         210         220         230         240
                    |           |           |           |           |           |
TRY1_HUMAN  LSQAKCEASYPGKITSNMFCVGFLEGGKDSCQGDSGGPVVCNG----QLQGVVSWGDGCA
CTRB_HUMAN  LSNAECKKSWGRRITDVMICAG--ASGVSSCMGDSGGPLVCQKDGAWTLVGIVSWGSDTC
            **:*:*: *:  :**. *:*.*    .* .** ******:**:       * *:****.. .
                    250         260
                    |           |
TRY1_HUMAN  QKNKPGVYTKVYNYVKWIKNTIAANS
CTRB_HUMAN  STSSPGVYARVTKLIPWVQKILAAN-
            ....****::*  :  : *::: :***
```

role is probably to hold the catalytic histidine in the correct orientation, which is also conserved between the two proteins (also shown in brown). The similarity between these two proteins cannot have arisen by chance. The only feasible explanation of the similarity in their structures is that they have been formed by **divergent evolution** from a common ancestral protein, also presumably a serine proteinase. The two proteins are said to be **homologues**, or **homologous** to one another. In fact, if one or other of these proteins is used for a BLAST search of the SwissProt database, dozens of related proteins will be found, all members of the serine proteinase family. Some of them are considerably larger than trypsin or chymotrypsin, and are found to have other recognizable structural **domains** or **motifs** associated with them. For example, prothrombin, which is the proenzyme of thrombin, has an N-terminal calcium-binding domain containing γ-carboxyglutamate (see Section 4.6.1), followed by two so-called **kringle domains** (so-called because kringle domains fold up into a shape very similar to that of a Danish pastry of the same name!), followed by the serine proteinase domain.

How has this happened? The likely explanation is that during copying of the DNA for cell division, a gene becomes duplicated; that is, an extra copy is produced. This extra copy is not required for the activities

Figure 6.5 Alignment of the sequences of human trypsinogen and chymotrypsinogen. The active site residues are shown in brown

Note that the sequences of trypsin and chymotrypsin are relatively distantly related. Not only have amino acid substitutions occurred since their divergence from a common ancestor, but there has also been deletion and addition of residues. This is the reason for the gaps introduced during alignment of the sequences (indicated by -) in Figure 6.5.

of the cell, and so can accumulate mutations until it acquires a new function. This process must have occurred many times to produce a large family of proteins such as the serine proteinases. In the case of multi-domain proteins like prothrombin, the serine proteinase gene is thought to have fused with copies of the genes for the calcium-binding unit and for the kringle domain to produce a composite gene for the new protein. Some cases are known where two or more copies of the same gene fused together to produce a protein in which the amino acid sequence repeats itself along the chain. Although these domains will have evolved so that the sequences are now different, their evolution from a common source may still be recognizable. For example, if the first third, the central section and the last third of the protein **serum albumin** are aligned, it can be seen that they were produced by this mechanism (Figure 6.6). The number of the position at which the same residue occurs in all three segments of the chain is small but significant; the gene triplication and fusion that gave rise to serum albumin was a very ancient event. Much greater similarity can be seen if the domains are compared pair-wise.

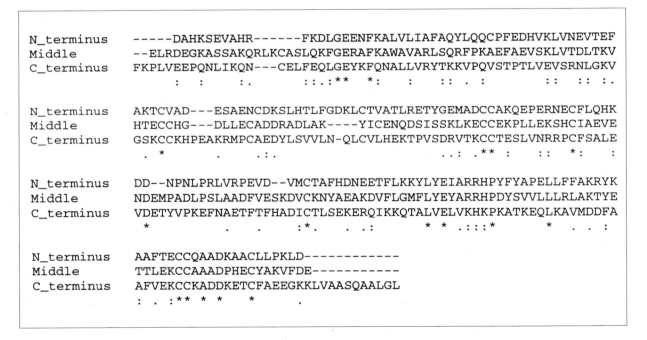

```
N_terminus   -----DAHKSEVAHR------FKDLGEENFKALVLIAFAQYLQQCPFEDHVKLVNEVTEF
Middle       --ELRDEGKASSAKQRLKCASLQKFGERAFKAWAVARLSQRFPKAEFAEVSKLVTDLTKV
C_terminus   FKPLVEEPQNLIKQN---CELFEQLGEYKFQNALLVRYTKKVPQVSTPTLVEVSRNLGKV
                  :    :     :.      ::.:**   *:     :     :: .  :        ::    :: :.

N_terminus   AKTCVAD---ESAENCDKSLHTLFGDKLCTVATLRETYGEMADCCAKQEPERNECFLQHK
Middle       HTECCHG---DLLECADDRADLAK----YICENQDSISSKLKECCEKPLLEKSHCIAEVE
C_terminus   GSKCCKHPEAKRMPCAEDYLSVVLN-QLCVLHEKTPVSDRVTKCCTESLVNRRPCFSALE
              .  *        .     .:.            ..:  .**  :     ::    *:      :

N_terminus   DD--NPNLPRLVRPEVD--VMCTAFHDNEETFLKKYLYEIARRHPYFYAPELLFFAKRYK
Middle       NDEMPADLPSLAADFVESKDVCKNYAEAKDVFLGMFLYEYARRHPDYSVVLLLRLAKTYE
C_terminus   VDETYVPKEFNAETFTFHADICTLSEKERQIKKQTALVELVKHKPKATKEQLKAVMDDFA
              *            .   .    :*.    . .:        *  *  .:::*       *    . . :

N_terminus   AAFTECCQAADKAACLLPKLD-----------
Middle       TTLEKCCAAADPHECYAKVFDE----------
C_terminus   AFVEKCCKADDKETCFAEEGKKLVAASQAALGL
             : . :** * *    *            .
```

Figure 6.6 Alignment of the three domains of human serum albumin (P02768)

In summary, then, it seems that the large number of complex proteins that we find in present day organisms have arisen by evolution from a much smaller set of simpler proteins by a process of gene multiplication. The extra copies then mutated until a new function was achieved, or fused with other genes to make multi-domain proteins. The evidence for these events is there to be seen if we carry out alignment of protein sequences.

6.6   Prediction of Three-dimensional Structures

The holy grail of the protein chemist is to be able to predict three-dimensional structures from amino acid sequences. The reason for this is that DNA sequencing is providing a vast number of protein sequences, but to understand how the proteins work we need to know their three-dimensional structures. Structure determination by X-ray diffraction is, however, a major undertaking, and the structures of only 2000–3000 different proteins are known. It would be an enormous advance if we could predict these structures rather than having to determine them experimentally.

We know that the instructions for protein folding are in some way encoded in the amino acid sequence (Section 5.6), but we do not know how that code works. One approach might be to assume that a protein folds into the conformation of lowest energy, and then to find that conformation by calculating the energies of all possible folds. There are several things wrong with this. Firstly, the number of folds that would have to be considered is simply too large. Secondly, although there are ways of calculating the energies of the folded states of proteins, they are necessarily imprecise – we do not know enough about the energies of interactions within proteins and of interactions with the solvent – and it is by no means certain that the structure of lowest energy could be found. Thirdly, although it is true that a protein folds to a state of lower energy than that of the unfolded form, there is no guarantee that the folded state is the one of lowest possible energy. Rather, the folded state might be the one that is achieved most rapidly; that is, folding might be under **kinetic** rather than **thermodynamic control**.

What, then, can be done? The only successful approach at the present time is **homology modelling**. We have seen in Section 6.5 that proteins fall into families, such as the serine proteinases, all the members of which have related amino acid sequences. The reason why changes in amino acid sequence can occur while still retaining the biological activity of the members of the family – the ability to hydrolyse proteins in the case of the serine proteinases – is that the changes are such as to keep the three-dimensional structures essentially unchanged, and to keep critical amino acid residues in the right place. Indeed, it turns out that three-dimensional structures are much more highly conserved than are amino acid sequences.

Consider, for example, the two structural cartoons in Figure 6.7. The one on the left is that of a monomer of triose phosphate isomerase (see Section 5.3.2). The one on the right is a monomer from the hexameric protein D-ribulose 5-phosphate 3-epimerase from potato (*Solanum tuberosum*, 1RPX[8]). These two proteins have very distantly related amino acid sequences (about 15% of identical residues). Nevertheless, their

At the time of writing, there were 17,869 entries in the Protein Data Bank at RCSB. Many of these, however, are variants of the same structure. For example, there were 779 entries for lysozyme.

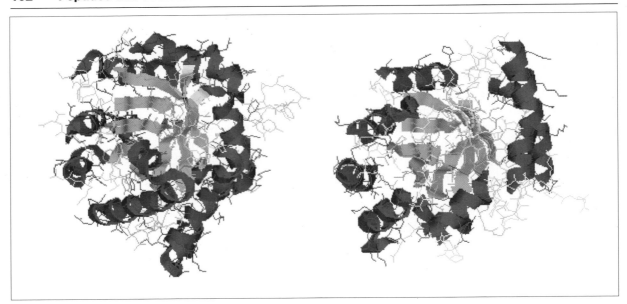

Figure 6.7 Cartoons of the three-dimensional structures of triose phosphate isomerase (*left*) and D-ribulose-5-phosphate 3-epimerase (*right*)

three-dimensional structures are related. They share the feature of an eight-stranded β-barrel core surrounded by helices. The number and arrangement of the helices differ somewhat, but the general pattern is similar. These two proteins, therefore, belong to the same structural family, and even though they have evolved to catalyse different types of reactions on different substrates (but note that the substrate is a sugar phosphate in both cases), their three-dimensional structures have remained very similar.

Homology modelling relies on the proposition that if two proteins can be shown to be homologous on the basis of sequence similarities, then it can be assumed that they share a common three-dimensional structure. To be safe in assigning structural similarity, it is usual to require that the degree of sequence identity should be at least 25% (although the case described above shows that this can be conservative). So suppose that we have a new protein whose three-dimensional structure we wish to predict. The first move is to carry out a database search to identify homologues; in particular, we require homologues for which three-dimensional structures have already been determined. If this search is unsuccessful, then the project cannot be proceeded with. If the search is successful, then the process is as follows.

- Align the amino acid sequences of the two proteins.
- Assign those stretches of amino acid sequence that are closely related as **structurally conserved regions (SCRs)**.
- Copy the atomic coordinates of the SCRs from the protein of known structure, and use them to construct a partial model of the unknown protein.

- Remove from the model any side chains that are different between the two proteins within the SCRs, and replace them with the correct side chains taken from a structure library.

  At this stage we will have a partial model of the unknown protein consisting of pieces of structure (the SCRs) separated by gaps. The stretches of sequence in these gaps are usually referred to as **loops**. To complete the model, the gaps have to be filled in. This can be done by **loop searching** as follows.

- Determine the geometry of the residues immediately before and after the loop.

- Go to a structure database and find a protein that contains a sequence of amino acids of the same length, with as nearly as possible the same amino acid sequence, and with the same geometry before and after it, as the loop we are trying to model.

- Take the atomic coordinates of that stretch of amino acids from the protein in the database, and assign them to the loop in the model.

- Again, replace any side chains that differ between the unknown protein and the loop from the protein in the database.

- Repeat the process for all of the loops.

At the end of this process we will have a complete model for the three-dimensional structure of the unknown protein. It will almost certainly contain some errors – incorrect bond lengths, distorted bond angles – because of the way in which it has been built. These distortions can be corrected by a process known as **molecular mechanics** in which all the atoms of the model are allowed to move by a small amount, the energy of the molecule is calculated, and the process repeated until an energy minimum is reached. This is only a very brief description of homology modelling. For a more detailed account, including on-line tutorials, see Martz.[9]

Homology modelling is obviously a computationally intensive process and requires specialized commercial software. It also requires the modeller to be make judgements at various stages during the process about such things as the appropriate choice of SCRs, the most suitable loop structure from a set of possible ones, and so on. For a protein chemist who does not have access to the software, and who simply wants to get a rough idea of what a new protein looks like, a fully automated structure prediction service called SWISS-MODEL is offered at the ExPASy site (http://www.expasy.org/swissmod/SWISS-MODEL.html). This comes with a health warning that the modelling procedure is fully automated, with no human intervention to optimize choices, and hence the models produced must be viewed with caution!

All that SWISS-MODEL requires as input is the amino acid sequence of the unknown protein. It then looks to see if there are any homologues with 30% or greater sequence identity and of known structure. If not, it reports the fact and closes. This occurred, for example, when the sequence of the _A. mellea_ nuclease (Section 6.4) was submitted. If the prediction is successful, the results are sent to the submitter by e-mail in a choice of formats, _e.g._ as a PDB file for displaying in RasMol.

6.7   Proteomics

Genomics is the study of the structures of the genomes of organisms. **Proteomics** is the study of which proteins are actually made in a given organism, or a in a particular cell type within an organism. That is, it is the study of how the genetic information is actually expressed. This is a topic of enormous current interest. For example, insights into diseases can be obtained by studying the differences in expression of proteins between normal and diseased cells.

The most powerful technique in proteomics is to separate the proteins from a cell extract by **two-dimensional electrophoresis,** and then to identify the proteins of interest by mass spectrometry. This technique will only be outlined here; for a review, see Andersen and Mann.[10]

The first step is the electrophoretic separation. This is carried out using polyacrylamide gels. In the first dimension, the proteins are run in a pH gradient. Each protein migrates to the point in the gel where it has no net charge (the isoelectric point; see Section 1.7) and then stops; this technique is called **isoelectric focusing.** The second dimension is then run in the presence of sodium dodecyl sulfate (see Section 4.2), where the proteins run according to their subunit relative molecular masses. Two-dimensional electrophoresis is capable of resolving mixtures containing more than 1000 different proteins and is very sensitive; proteins can be detected at the femtomol (10^{-15} mol) level. There are now databases with collections of electrophoresis patterns produced by different cell types, and software is available for comparison of a new map with those in the database. An example of two-dimensional electrophoresis of a cell extract is shown in Figure 6.8; it was obtained from the SWISS-2DPAGE database.

Once a map of the protein mixture is obtained, it remains to identify the protein of interest. Sometimes this can be done by searching a database which lists proteins by M_r and isoelectric point, both of which can be determined from the position of a protein in the map. There are problems with this. Firstly, there may be several proteins present with similar values for these parameters. More seriously, many proteins are subject to post-translational modifications which can alter both the M_r and the pI values. Mass spectrometry provides an answer to this problem (see Section 4.6.2). First the protein is extracted from the gel, and then it is digested with trypsin. This produces a set of peptides, the masses of which can be determined by mass spectrometry. A complete set cannot usually be obtained, but that is not a problem. These peptide masses are then used to search a database which contains, for every known protein, the masses of the tryptic peptides that would be produced by digestion with trypsin. What is being looked for is a protein in the database which produces peptides with the same masses as those

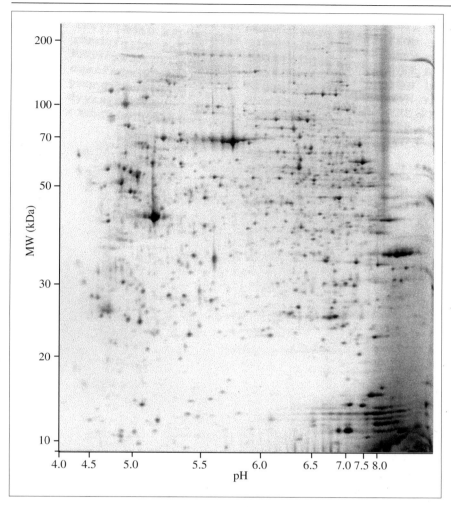

Figure 6.8 Two-dimensional electrophoresis pattern of proteins from a human melanoma. Isoelectric focusing was carried out in the horizontal direction and SDS electrophoresis in the vertical direction. The image was taken from the SWISS-2DPAGE database[11]

used in the search. Generally, if all the masses submitted match with those from the protein in the database, and if at least 20% of the sequence of the protein is covered, this is considered sufficient to provide unambiguous identification of the protein. This is a good example of how the power of modern analytical methods is combined with bioinformatics to yield information of great value, not only for understanding how the genome is expressed, but also for understanding disease processes.

Worked Problem 6.7

Q A protein was isolated from the bacterium *Helicobacter pylori* and run on SDS-PAGE. The protein band was extracted, digested with trypsin, and subjected to mass spectrometry. The following is a list of *m/z* values for the [M + H]⁺ ions of peptides from the digest: 692.37, 919.5, 971.53, 990.5, 1095.5, 1122.5, 1132.6, 1268.6, 1443.6, 1517.8, 1594.9, 1657.9, 2339.2, 2534.2. Use this information to identify the protein. (The protein isolation was done by Dr Ravi Nookala, and the digestion and mass spectrometry by Dr Clive Slaughter).

A The peptide masses were submitted to the program **PeptIdent** at the ExPASy server (http://ca.expasy.org/tools/peptident.html). The database to be searched was chosen as SwissProt, and the species to be searched selected as Bacteria. No values were given for p*I* and MW. All other values were left as defaults. The peptide *m/z* values were entered and PeptIdent run.

The results, sent by e-mail, reported that 12 proteins were found in which four or more of the peptides matched, but the only protein for which all peptides matched was the enzyme **catalase** from *H. pylori*. A list of the sequences of the matched peptides was given (not shown), followed by the complete sequence of the protein with the relevant peptides shown in brown capitals (Figure 6.9; note that all peptides are preceded by, and terminate in, Lys or Arg, as required for a tryptic digest). The matched peptides covered 26.1% of the total protein. The extent of the coverage, and the complete matching of the peptides, confirmed that the protein was indeed catalase. The output also gave the M_r of the protein as 58,629.14 and the p*I* as 8.70. Interestingly, it also reported the computer time used for the complete analysis as 14.04 seconds!

Figure 6.9 The amino acid sequence of catalase from *H. pylori* showing in brown the tryptic peptides whose masses were used to identify the protein

```
  1  mvnkdvkqtt  afgapvwddn  nvitagprGP  VLLQSTWFLE  KLAAFDReri  pervvhakgs   60
 61  gaygtftvtk  ditkytkaki  fskvgkktec  ffrfstvage  rgsadavrdp  rgfamkYYTE  120
121  EGNWDLVGNN  TPVFFIRdai  kFPDFIHTQK  rdpqtnlpnh  dmvwdfwsnv  peslyqvtwv  180
181  msdrgipksf  rhmdgfgsht  fslinakger  fwvkfhfhtm  qgvkHLTNEE  AAEVRkydpd  240
241  snqrDLFNAI  ARgdfpkwkl  siqvmpeeda  kkyrFHPFDV  TKiwylqdyp  lmevgiveln  300
301  knpenyfaev  eqaafspanv  vpgigyspdr  mlqgrLFSYG  DTHRyrLGVN  YPQIPVNKPR  360
361  cpfhsssrdg  ymqngyygsl  qnytpsslpg  ykedksardp  kFNLAHIEKE  FEVWNWDYRa  420
421  ddsdyytqpg  dyyrslpade  kerLHDTIGE  SLAHVTHKei  vdkqlehfkk  adpkyaegvk  480
481  kalekhqkmm  kdmhgkdmhh  tkkkk
```

Summary of Key Points

- Databases of protein sequence and structure are electronic repositories of validated and annotated data, and contain computer programs to extract and analyse those data.

- Bioinformatics is the science of the storage, retrieval and analysis of information on biomolecular structure.

- Models of protein structures can be drawn using molecular graphics programs such as RasMol. The input for such programs is the type, and the Cartesian coordinates, of all atoms in the molecule (except hydrogen).

- The biological function of a newly discovered protein can often be found by comparison of its sequence with those in a database of known proteins. This is because proteins with related functions have related amino acid sequences.

- Protein sequence comparisons also throw light on the processes of evolution. The degree of sequence difference between two proteins that have the same function in two different organisms is greater the further apart the organisms are on the evolutionary time scale.

- Families of proteins with related functions have arisen by a process of gene multiplication and divergent evolution of the multiple copies. Some proteins have arisen by the fusion of two or more genes to produce proteins with multiple domains.

- In proteins that have been produced by evolution from a common ancestor, the three-dimensional structure is conserved. This means that the three-dimensional structure of an unknown protein can be predicted by homology modelling if a homologue can be identified whose structure is already known.

- Proteomics is the identification of the proteins that are produced in a particular cell type. A widely used technique in this field involves separation of a protein extract by two-dimensional electrophoresis. This is followed by digestion of the individual proteins with trypsin, and analysis of the product peptides by mass spectrometry. The peptide masses obtained are then used to search a database containing the masses of tryptic peptides derived from known proteins.

Problems

6.1. Locate the three-dimensional structure of melittin in the RCSB database and download it. Examine the downloaded file. For how many polypeptide chains are the coordinates reported? Produce a model of a single chain in wireframe using RasMol, and with the colour system adopted in this book. Comment on the secondary structure of the peptide.

6.2. Locate the amino acid sequence of the cytochrome *c* from cauliflower in the SwissProt database. Use CLUSTAL W to align the sequence with that of the same protein from humans. Copy and display the sequence alignment. From the percentage identity between the proteins, and using equation (6.1), calculate the elapsed time since humans and cauliflowers last shared a common ancestor.

6.3. A protein was isolated from the fungus *A. mellea* and found to have the N-terminal sequence ISYNGWTSSRQTTLVSAAAA WQTYAQ.[12] Carry out a BLAST search of the SwissProt database in an attempt to identify what type of activity the protein might have.

6.4. Download from the RCSB database the structure files 2CHA[13] (bovine chymotrypsin) and 2BZA[14] (bovine trypsin) (the identifiers are given because if either trypsin or chymotrypsin is entered as a search term, there are a very large number of structures to choose from). In both cases, view the molecule as a backbone model. Select residues 57, 102 and 195 (the active site residues) and draw them in spacefill. Orientate the molecule so that the active site residues appear to be in the front and in the middle of the view, with Ser-195 at the bottom. Compare the structures of the two molecules.

6.5. Produce a stereo view of bovine chymotrypsin in the same orientation as in Problem 6.4.

6.6. Two-dimensional electrophoresis was carried out on a human liver extract. One protein, with a subunit M_r of about 17,000, was eluted from the gel, digested with trypsin, and the product peptides analysed by mass spectrometry. The following are the m/z values of the $[M + H]^+$ ions of the peptides identified: 2116.8, 1251.6, 1023.5, 768.3, 389.3. Use the program PeptIdent at SBI to identify this protein. When asked for the database to be searched, choose SwissProt. When asked for the species to be searched, choose *Homo sapiens* (human). Leave all other entries as the defaults. Enter the peptide masses, your e-mail address, and Run. After you have identified the protein, look at its SwissProt entry and find out what sort of activity it has.

References

1. C. Gibas and P. Jambeck, *Developing Bioinformatics Computer Skills*, O'Reilly, Sebastopol, 2001.
2. S. F. Altschul, W. Gish, W. Miller, E. W. Myers and D. J. Lipman, *J. Mol. Biol.*, 1990, **215**, 403.
3. V. Healy, S. Doonan and T. V. McCarthy, *Biochem J.*, 1999, **339**, 713.
4. J. D. Thompson, D. G. Higgins and T. J. Gibson, *Nucleic Acids Res.*, 1994, **22**, 4673.
5. L. Pathy, *Protein Evolution*, Blackwell, Oxford, 1999.
6. M. O. Dayhoff, *Atlas of Protein Sequence and Structure*, National Biomedical Research Foundation, Washington, 1972.
7. M. Kimura, *The Neutral Theory of Molecular Evolution*, Cambridge University Press, Cambridge, 1983.
8. J. Kopp, S. Kopriva, K. H. Suss and G. E. Schultz, *J. Mol. Biol.*, 1999, **287**, 761.
9. E. Martz, *Homology Modelling for Beginners*, at http://www.umass.edu/microbio/chime/explorer/homolmod.htm#intros, 2001.
10. J. S. Andersen and M. Mann, *FEBS Lett.*, 2000, **480**, 25.
11. C. Hoogland, J.-C. Sanchez, L. Touella, P.-A. Binz, A. Bairoch, D. F. Hochstrasser and R. D. Appel, *Nucleic Acids Res.*, 2000, **28**, 286.
12. V. Healy, J. O'Connell, T. V. McCarthy and S. Doonan, *Biochem. Biophys. Res. Commun.*, 1999, **262**, 60.
13. J. J. Birktoft and D. M. Blow, *J. Mol. Biol.*, 1972, **68**, 187.
14. J. D. Bartunik, L. J. Summers and H. H. Bartsch, *J. Mol. Biol.*, 1989, **210**, 813.

Further Reading

T. K. Attwood and D. J. Parry-Smith, *Introduction to Bioinformatics*, Pearson, Harlow, 1999.

D. Higgins and W. Taylor, *Bioinformatics: Sequence, Structure and Databanks*, Oxford University Press, Oxford, 2000.

Answers to Problems

Chapter 1

1.1. The two stereogenic centres are shown in **1** and **2**, with arrows indicating the direction of priority of the attached groups. In **1** the configuration is *S* and in **2** it is *R*. So L-threonine has the configuration (2*S*,3*R*).

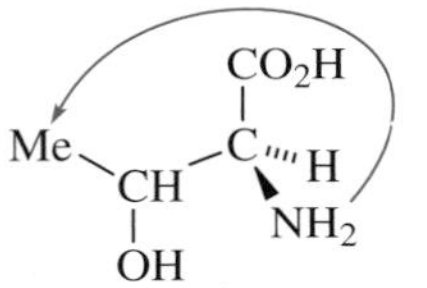

1

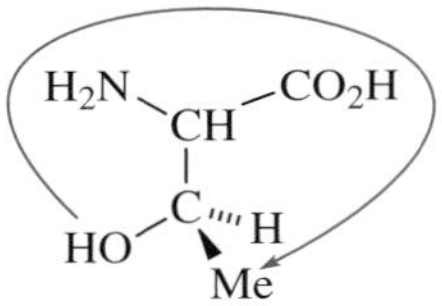

2

1.2. The amide group is a resonance hybrid of the structures shown:

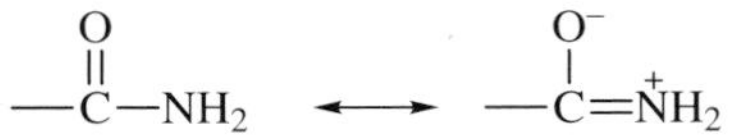

Protonation of the amide nitrogen would remove the possibility of resonance stabilization.

1.3. The two possible structures are shown in **3** and **4**. Compound **3** is Trp-Asn, or WN, whereas **4** is Asn-Trp or NW.

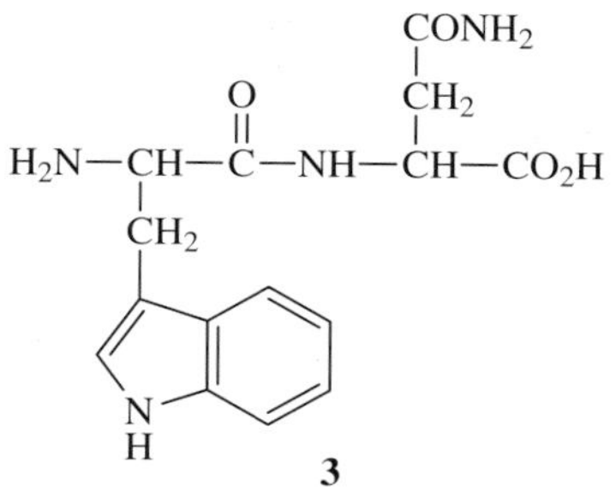

1.4. The structure is shown in **5**.

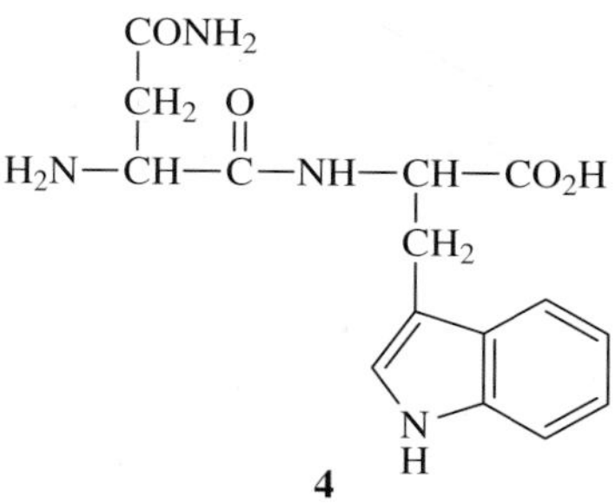

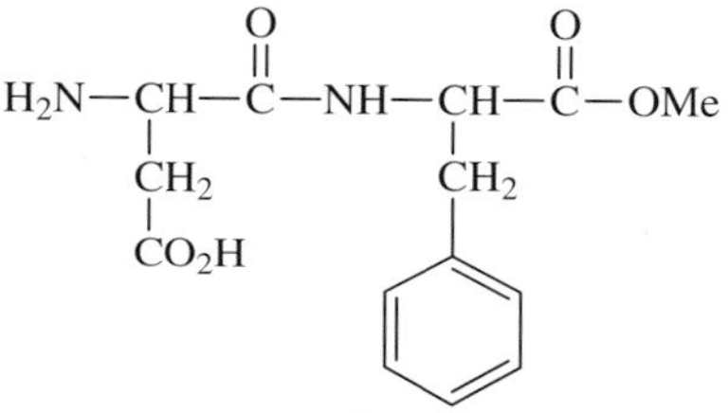

1.5. The point here is that the neither of the side chains is basic, so the only ionizable groups are the primary amino and carboxylic acid groups. Hence at pH 2 the charge will be +1, and at pH 12 it will be −1.

1.6. The N-terminal residue could be any of 4, the second residue any of 3, and the third either of 2. Hence there are 24 possibilities. They are:
DWIF; DWFI; DIWF; DIFW; DFIW; DFWI; WDIF; WDFI; WIDF; WIFD; WFID; WFDI; IWDF; IWFD; IDFW; IDWF; IFDW; IFWD; FWDI; FWID; FDIW; FDWI; FIDW; FIWD

1.7. The composition is $A_4D_1F_2G_2K_1P_1S_1W_1$. The sequence in three-letter abbreviations is:

Ala-Asp-Ala-Gly-Phe-Trp-Lys-Phe-Ala-Ala-Gly-Pro-Ser

1.8. Each position in the sequence can be occupied by any one of 20 residues, and therefore there are 20^{50} possibilities. This is 1.126×10^{65}. The average residue $M_r = 110$. Hence the average protein $M_r = (50 \times 110) + 18 = 5518$ (the extra 18 comes from the H on the N-terminus and the OH on the C-terminus). The mass of a single molecule of each of these proteins would be $5518/(6 \times 10^{23}) = 9.197 \times 10^{-21}$ g. Hence the total mass of one molecule of each of them would be 1.036×10^{45} g. [Note that the mass of the Earth is 5.98×10^{27} g]

Chapter 2

2.1. An N-protected (and, if necessary, side chain-protected), carboxyl-activated derivative is required, *i.e.* **6**. The product of the reaction is structure **7**, with the newly introduced residue in brown.

2.2. The yield of the 13-mer will be less than that of the 12-mer and so will determine the overall yield. The yield of the 13-mer will be $100 \times 0.95^{12} = 54.0\%$. Coupling to the 12-mer occurs with a yield of 95%. Hence the overall yield will be 51.3%.

2.3. The completed scheme is given below (assuming that the last residue does not require side chain protection). It would remain to liberate the tripeptide from the solid support and to de-protect the N-terminus.

2.4. The scheme is as follows. The protecting groups are shown in brown. The *t*-BOC group is not removed by piperidine.

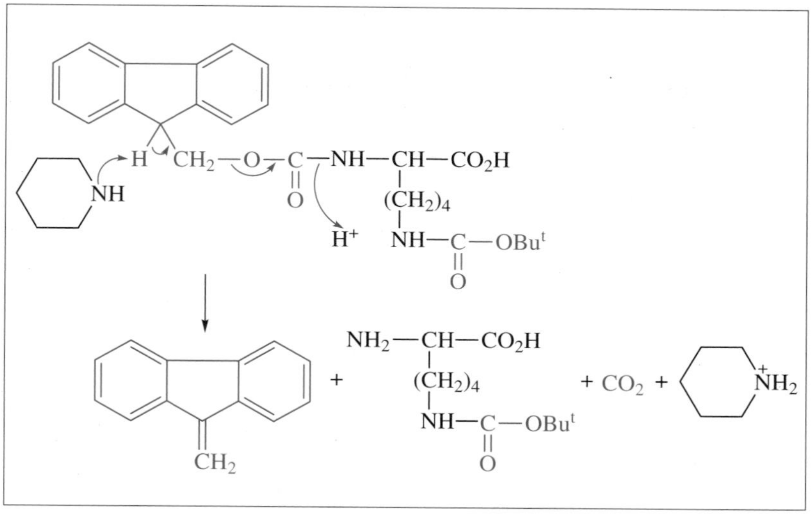

2.5. For the *t*-Boc/benzyl strategy the derivative would be N^α-*t*-Boc-aspartic acid benzyl ester (**8**), and for the Fmoc/*t*-butyl strategy it would be N^α-Fmoc-aspartic acid *t*-butyl ester (**9**).

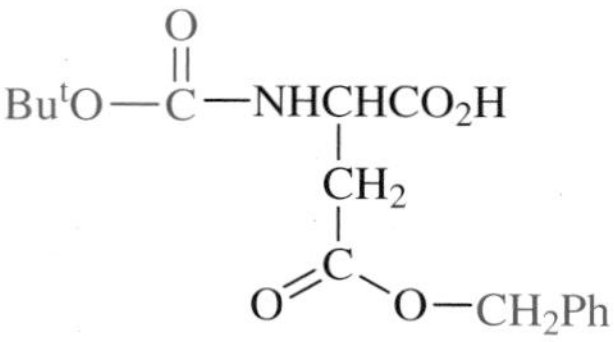

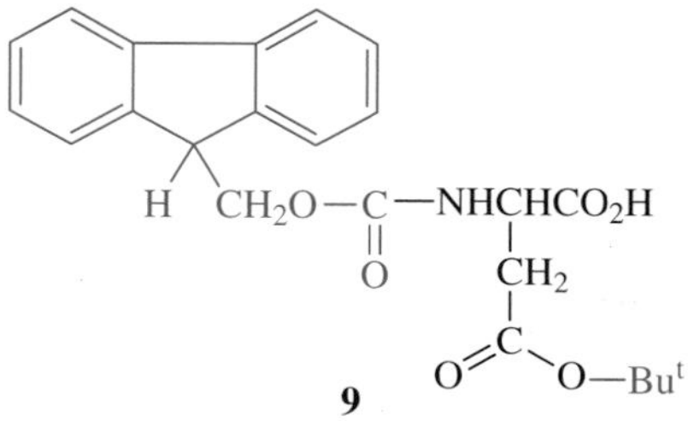

2.6. The reaction sequence is shown here, in which the cyclohexyl groups of the DCC are represented by R^1:

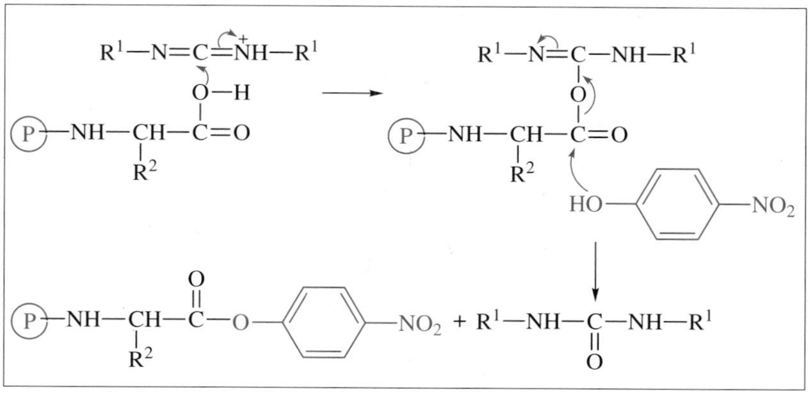

2.7. DCC coupling would be used as shown in the following scheme. The initial reaction of Fmoc-Gly (shown with the Fmoc group abbreviated for simplicity) with DCC is exactly as shown in Problem 2.6.

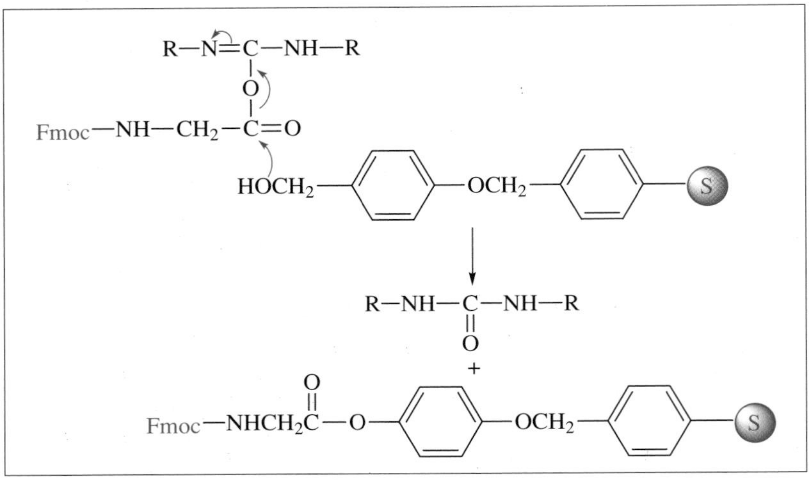

2.8. The synthesis is shown in the following scheme. For simplicity, the Wang resin and the protecting groups are shown in abbreviated form.

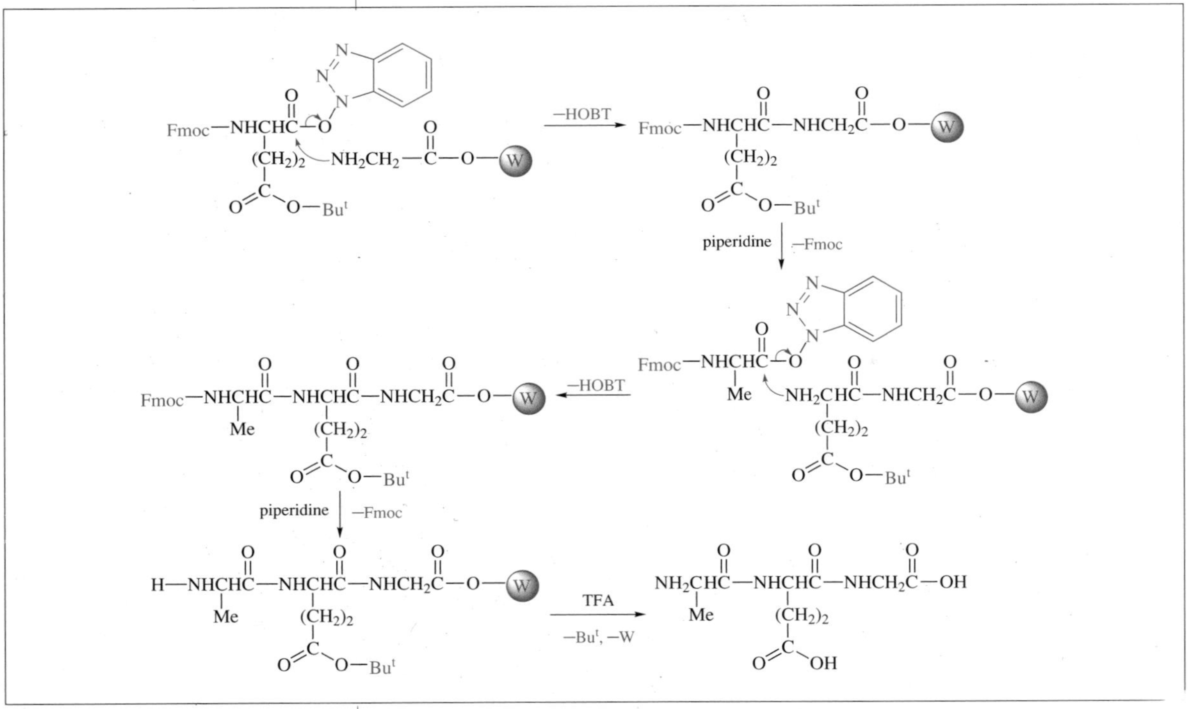

Chapter 3

3.1. The buffer will contain 0.075 mol dm⁻³ $H_2PO_4^-$ (proton donor) and 0.025 mol dm⁻³ HPO_4^{2-} (proton acceptor). Thus from equation (3.2):

$$pH = 7.20 + \log \frac{0.025}{0.075} = 6.72$$

3.2. Suppose that HCl is added to a concentration of x mol dm⁻³. The concentration of free base remaining will be $(0.10 - x)$ mol dm⁻³, and that of the protonated base will be x mol dm⁻³. Hence, from equation (3.2):

$$7.50 = 8.00 + \log \frac{0.10 - x}{x}$$

Solving this gives $x = 0.076$ mol dm⁻³, and since the volume is 1 dm³ we must add 0.076 mol of HCl.

3.3. The relationship between relative centrifugal field (RCF), radius r and angular velocity ω is:

$$RCF = r\omega^2 \times 1.119 \times 10^{-5} \, g$$

Putting RCF = 10,000 g and $r = 25$ in this equation, and solving for ω, gives 6000 rpm.

3.4. Call the proteins with pI values of 6.0, 7.0 and 8.0 as A, B and C, respectively. One way of doing it would be to equilibrate the column at a pH of, say, 9.0, at which all of the proteins would be negatively charged. On application to the column, all three proteins would be bound to the resin. Application of a salt gradient would then elute the proteins in the order C, followed by B, followed by A.

One problem with this is that protein A might be quite highly charged at pH 9.0 (three pH units above its pI) and difficult to elute. It would be better to choose a pH, say 7.5, at which C would be positively charged, and the other two negative. In this case, C will pass straight through the column. A and B will bind, and could be eluted by application of a salt gradient. B will elute first, followed by A. Obviously a pH less than 7.0 could not be used, because both B and C would pass straight through under those conditions.

It is not possible to predict how these three proteins will behave on SDS-PAGE. Mobility on SDS-PAGE depends on the protein's (subunit) relative molecular mass, not on its charge.

3.5. The obvious technique to use is ion exchange chromatography with a negatively charged resin such as CM-cellulose. At a pH of, say, 7.0 the genetically engineered protein would be strongly positively charged because of the tail of lysine or arginine residues. Most other proteins in the mixture would be either negatively charged, or weakly positively charged at this pH. After application of the protein mixture to the column and elution of the unbound proteins, the protein of interest would be eluted by application of a salt gradient.

There is, of course, the possibility that the protein mixture contained another strongly positively charged protein, arising from an abnormally high internal content of basic residues. A useful trick in this case is to remove the poly-Lys or poly-Arg tail from the partially purified protein using carboxypeptidase B (this is an exoproteinase specific for Lys and Arg). If chromatography is repeated under the same conditions as before, the protein of interest will no longer bind whereas the impurity, which will be unaffected by the proteinase, will bind as before. The protein of interest, which now passes straight through the column, should be pure.

3.6. (a) The presence of a salt, such as ammonium sulfate, in a protein solution promotes binding to hydrophobic surfaces. Hence it is an advantage in hydrophobic interaction chromatography (see Section 3.3.4).

(b) The two methods used to remove salt from a protein solution are dialysis and gel permeation chromatography (see Section 3.3.6).

3.7. The completed purification table is shown below.

Fraction	Volume (cm³)	Protein concentration (mg/cm³)	Total protein (mg)	Activity (U/cm³)	Total activity (U)	Specific activity (U/mg)	Purification factor	Overall yield (%)
Homogenate	1000	42.5	42,500	0.9	900	0.0212	1	(100)
45–65% (NH₄)SO₄	120	134.0	16,080	6.8	816	0.0507	2.4	91
Hydrophobic chromatography	230	3.7	851	0.8	184	0.216	10.2	20
DEAE-Sephadex	55	0.92	50.6	2.4	132	2.61	123	15
CM-Sephadex	15	0.48	7.2	7.2	108	15.0	708	12

This is a rather poor purification schedule because of the low over-all yield of only 12%. The problem was with the hydrophobic chromatography, where a step yield of only 23% was achieved. It is significant that the volume of the active fraction at this step was large (230 cm³). It looks as if the protein of interest was too strongly bound to the hydrophobic support and leaked off slowly when the salt concentration was decreased. This step should be replaced.

Chapter 4

4.1. To allow for the fact that the printed size of Figure 4.8 will differ from that of the original gel, the results have been analysed in terms of the **relative mobility** of the protein bands; that is, the distance moved divided by the total length of the gel. Log M_r values are plotted against the relative mobilities in the figure below. Excel was used to plot the results and to calculate the equation of the trend line (shown on the plot). The relative mobility of the unknown was 0.0988. This leads to an M_r of 39,000.

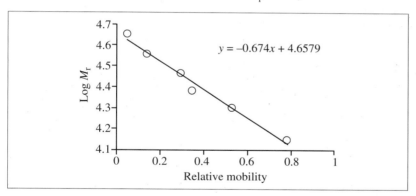

$$y = -0.674x + 4.6579$$

This result is likely to be imprecise. The unknown has not migrated far into the gel. The experiment should be repeated with a gel of lower degree of cross-linking, and with more standards in the high M_r range.

4.2. The likely explanation of these observations is that the protein is a tetramer with two identical subunits of $M_r = 43,000$ and two with $M_r = 40,000$. This is consistent with the M_r of 166,000 observed on gel permeation chromatography. The protein cannot consist of a dimer with each monomer consisting of two chains linked by disulfide bridges, because the SDS-PAGE was run in the absence of mercaptoethanol. The fact that only two different N-terminal residues were detected provides further evidence that the protein consists of only two different polypeptide chains.

4.3. Putting the values given into equation (4.1) leads to:

$$M = \frac{4.41 \times 10^{-13}}{6.30 \times 10^{-11}} \times \frac{8.314 \times 293}{(1 - 0.75 \times 1.01)} = 70.3 \text{ kg mol}^{-1}$$

[The value of M_r from the amino acid sequence is 64,600. It is likely that the discrepancy is due to inaccuracy in the value of the partial specific volume. Note that haemoglobin is a tetramer of two different subunits. The adult forms of these subunits are called α and β, so haemoglobin has the quaternary structure $\alpha_2\beta_2$. The subunits are structurally related, and both are related to the monomeric oxygen-binding protein myoglobin. They form part of the family of the globins (see Section 6.5 for a treatment of protein evolution).]

4.4. The four sets of values are plotted in the figure below. It is clear that Leu is liberated first (open squares) and must be the C-terminal residue. Similarly, Phe is liberated next (open circles). The problem is with Gly and Trp, which are liberated at essentially the same rate. To interpret these results it is necessary to know that Gly is a poor substrate for carboxypeptidase A, whereas Trp is a good substrate. So the interpretation is that Gly precedes Phe in the sequence, and is liberated slowly. Gly is preceded by Trp, which is liberated as soon as the Gly is removed. Hence the sequence is Trp-Gly-Phe-Leu.

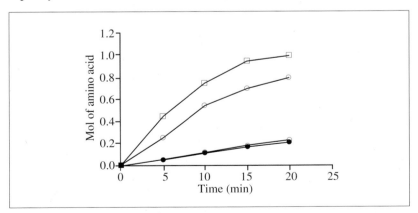

4.5. The cyanogen bromide peptide has an N-terminal sequence the same as the C-terminus of the second tryptic peptide, and a C-terminus the same as the N-terminus of the first tryptic peptide. Hence these two can be overlapped to give TMADRILSMR. The N-terminus of the thermolytic peptide is the same as the C-terminus of this composite peptide, and its C-terminus is the same as the N-terminus of the third tryptic peptide. This leads to TMADRILSMRSELR for the partial sequence of the protein. This may be clearer in the following figure, where the residues are given in the three-letter code, and the position of each of the peptides is indicated. [Note that these peptides were obtained as part of a project to sequence a protein which turned out to contain 412 amino acid residues.]

```
Thr-Met-Ala-Asp-Arg-Ile-Leu-Ser-Met-Arg-Ser-Glu-Leu-Arg
< ------Tr2------ > < -----Tr1------- > < ----Tr3---- >
       < --------CNBr---------- >
                    < -------Th--------- >
```

4.6. Taking the differences between the m/z values of successive sequence ions gives 158, 127, 129, 85 and 232. These are identical with those in the Worked Problem with the exception of the third one, which in the present case is 129 compared with 115. This difference of 14 units in mass suggests that the residues differ by a CH_2 unit, which in turn suggests that the residue in the present peptide is Thr rather than Ser. In fact, no other internal residue has this mass value. The sequence is, therefore, AcSer-Leu-**Thr**-Ala-Lys.

4.7. Refer to Table 1.1 for the M_r values of the amino acids. The ion at $m/z = 221$ is too large to be any of the amino acids and so must correspond to $b_1 + b_2$. It should, therefore, be the sum of two amino acid M_r values minus 35 (two OH groups and one H). The only pair that fits is Gly (75) and Tyr (181), but it is not possible to say which is N-terminal. After that it is straightforward. The mass difference between $b_1 + b_2$ and b_3 is 57, which corresponds to Gly (75 – 18). That between b_3 and b_4 is 147, which corresponds to Phe (165 – 18). That between b_4 and b_5 is 113. This could be either Leu or Ile (131 – 18). Hence the sequence that is derived is either Gly-Tyr-Gly-Phe-Leu/Ile, or Tyr-Gly-Gly-Phe-Leu/Ile. Analysis of the y-type ions confirms the second of these possibilities.

4.8. The extra mass corresponds to a t-Boc group (Bu^{t}OCONH- compared with NH$_2$-), so it is likely that the N-terminal blocking group was only partially removed.

Chapter 5

5.1. The structure is shown in **10**. Note how the parallel arrangement of the chains causes the hydrogen bonds to incline away from the axis of the sheet.

10

5.2. By "rise" is meant the distance up the helix from a particular atom in one residue, say C_α, to the same atom in the next residue. Since there are 3.6 residues per turn, the rise is $3.6 \times 1.5 = 5.4$ Å. Compare collagen, where the rise is 9.6 Å.

5.3. In the right-handed α-helix made from L-amino acids, the side chains point outwards from the helix (see Figure 5.4). If the residues had the D-configuration, the side chains would be inside the helix, but there is no room for them there. D-Amino acids form a **left-handed α-helix**

5.4. $\Delta S_p = -100 \times 8.314 \times \ln 8 = -1730$ J mol^{-1} K^{-1}. Hence $-T\Delta S_p = 516$ kJ mol^{-1} K^{-1}. Compare this with the value of 444 kJ mol^{-1} K^{-1} obtained with $n = 6$. This shows that the value of ΔS_p is relatively insensitive to the assumed value of n, and gives confidence in the analysis of the factors responsible for folding given in Section 5.6.

5.5. Determination of the ratio b/a at a urea concentration of 4 mol dm^{-3} gives a value of 0.7. This is the fraction, α, of protein unfolded. From equation (5.4) this leads to an equilibrium constant

$K = 2.33$. The standard free energy change is then given by equation (5.5) as:

$$\Delta G° = -8.314 \times 298 \times \ln 2.33 = -2100 \text{ J mol}^{-1} \text{ or } -2.1 \text{ kJ mol}^{-1}$$

Chapter 6

6.1. The identifier for melittin is 2MLT (T. C. Terwilliger and D. Eisenberg, *J. Biol. Chem.* 1982, **257**, 6010). Under REMARK 4 in the file it is reported that melittin is a tetramer, but that the structures of only two of the chains (A and B) are included in the entry. The figure below shows the structure of the A chain coloured as usual, with wireframe thickness of 50, and restricted to chain A (command *restrict *A*). Clicking on *Cartoons* shows that the chain is essentially α-helical with a short section (two residues) where the helix is interupted.

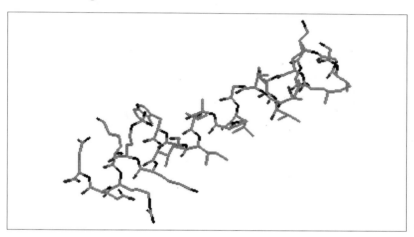

6.2. The identifier of the protein in SwissProt is P00050. That of the human protein is P00001. Using these as inputs for CLUSTAL W yields the sequence alignment shown (after editing) in the figure below. The output reports that the sequences show 68 identities in 112 residues, but note that the cauliflower protein is seven residues longer than that from humans. Restricting the comparison to the 104 residues of the human protein, the identity is 68 in 104; that is, 65.4% identical residues, or 34.6% different.

In the terminology used by Kimura, the fraction of sequence differences $p_d = 0.346$. So from equation (6.1), $K_{aa} = 0.462$, showing that 46.2% of changes have accumulated, 23.1% in each sequence. This gives the elapsed time as 494 million years. [Note that if the

```
                     10        20        30        40        50        60
                      |         |         |         |         |         |
CYC_BRAOL   ASFDEAPPGNSKAGEKIFKTKCAQCHTVDKGAGHKQGPNLNGLFGRQSGTTAGYSYSAAN
CYC_HUMAN   -------GDVEKGKKIFIMKCSQCHTVEKGGKHKTGPNLHGLFGRKTGQAPGYSYTAAN
            *: : *:***  **:*****:**. ** ****:*****::*  :.****:***
                     70        80        90       100       110
                      |         |         |         |         |
CYC_BRAOL   KNKAVEWEEKTLYDYLLNPKKYIPGTKMVFPGLKKPQDRADLIAYLKEATA-
CYC_HUMAN   KNKGIIWGEDTLMEYLENPKKYIPGTKMIFVGIKKKEERADLIAYLKKATNE
            ***.: * *.** :** **********:* *:** ::*********:**
```

elapsed time is calculated ignoring the possibility of multiple replacements, then the erroneous result of 390 million years is obtained.]

6.3. The essential part of the output from BLAST is shown below. Only one related protein was identified, with 52% identity in 23 residues. This protein was an endopeptidase (endoproteinase) from the fungus *Grifola frondosa*. The enzyme requires a metal ion (Zn^{2+}) for its activity, and so is called a **metalloendopeptidase**. It cleaves peptides and proteins on the N-terminal side of lysine residues. Because of the high degree of similarity between the sequences, and also because the similarity with the *G. frondosa* protein is also at its N-terminus, the match is highly significant, and it can be concluded that the *A. mellea* protein is also a lysine-specific metalloendopeptidase.

```
>sp|P81054|PLMP GRIFR Peptidyl-LYS metalloendopeptidase (EC
        3.4.24.20) (MEP) (GFMEP).[Grifola frondosa]
        Length = 167

 Score = 28.1 bits (61), Expect = 3.3
 Identities = 12/23 (52%), Positives = 18/23 (78%)

Query: 2   SYNGWTSSRQTTLVSAAAAWQTY 24
           +YNG +SS Q+ L +AA+A Q+Y
Sbjct: 1   TYNGCSSSEQSALAAAASAAQSY 23
```

6.4. The two models for trypsin and chymotrypsin are shown below. Comparing the two proteins, it is immediately apparent that their three-dimensional structures are very similar. This is what would be expected from the fact that their amino acid sequences are related (see Figure 6.5 for an alignment of the sequences of the human zymogens).

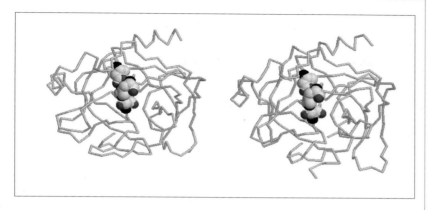

6.5. The stereo model is shown below. If you view the model by naked eye stereopsis, it is apparent that the active site residues are not at the front, as would appear from the figure in Answer 6.4, but rather are in the centre of the molecule.

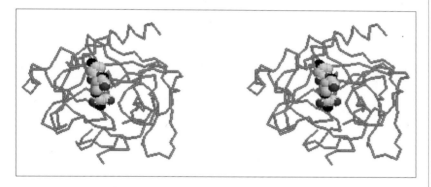

6.6. An edited version of the printout from PeptIdent is shown below. One protein (P00740) was found for which all five peptide masses matched, and which had an M_r of 16,488. The five peptides covered 33% of the sequence of the protein, as shown in brown in the lower part of the figure. A second protein (Q13315) matched four of the peptides, but the M_r of this protein was 350,644 and so it could be ruled out.

Clicking on the hyperlink for protein P00740 leads to a description of the protein. It is the light chain of an enzyme called **coagulation factor IX** (also known as **Christmas factor**) and is involved in the blood clotting process. Its substrate is coagulation factor X in which it hydrolyses a single Arg-Ile bond. It is a serine proteinase.

2 matches found.

Score	# peptide matches	AC	ID	Description	pI	Mw
1.00	5	<u>P00740</u>	FA9_HUMAN_1	CHAIN 1: FACTOR IXA (ACTIVE FORM) LIGHT CHAIN. - Homo sapiens (Human).	4.65	16488.24
0.80	4	<u>Q13315</u>	ATM_HUMAN	Serine-protein kinase ATM (EC 2.7.1.37) (Ataxia telangiectasia mutated) (A-T, mutated). - Homo sapiens (Human).	6.34	350644.35

33.1% of sequence covered:

```
          1          11         21         31         41         51
          |          |          |          |          |          |
     1                                               ynsg kleefvqgnl   60
    61 erECMEEKcs feearEVFEN TERttefwkQ YVDGDQCESN PCLNGGSCKd dinsyecwcp  120
   121 fgfegkNCEL DVTCNIKngr ceqfcknsad nkvvcscteg yrlaenqksc epavpfpcgr  180
   181 vsvsqtskLT R
```

Subject Index